PROJECT AIR FORCE

T0302804

Support to the Air Force Installation and Mission Support Center

Enabling AFIMSC's Role in Agile Combat Support Planning, Execution, Monitoring, and Control

Patrick Mills, Robert S. Tripp, James A. Leftwich, John G. Drew,
Jerry M. Sollinger, Robert G. DeFeo

Prepared for the United States Air Force

For more information on this publication, visit www.rand.org/t/RR1555

Library of Congress Cataloging-in-Publication Data is available for this publication.

ISBN: 978-0-8330-9655-5

Published by the RAND Corporation, Santa Monica, Calif.

© Copyright 2017 RAND Corporation

RAND® is a registered trademark.

Support RAND
Make a tax-deductible charitable contribution at
www.rand.org/giving/contribute

www.rand.org

Preface

The U.S. Air Force is in the midst of transforming the way that agile combat support (ACS) is managed to support training, steady state operations, and contingency operations. The reorganization of Air Force Materiel Command into the five-center construct, including the Air Force Life Cycle Management Center and the Air Force Sustainment Center, was one significant milestone in the transformation. Another is the establishment of the Air Force Installation and Mission Support Center (AFIMSC), guided by Headquarters, U.S. Air Force, Program Action Directive 14-04.

This report provides a strategic view of the analytical capabilities that are needed by AFIMSC to relate ACS resource levels and process performances to operationally relevant metrics. The report summarizes insights from prior RAND Project AIR FORCE (PAF) research and proposes a theoretical model and analytical framework that can be used to improve AFIMSC planning and execution activities. These analytical capabilities are essential if AFIMSC is to develop options for meeting demands for ACS resources and to relate how each option affects effectiveness, costs, and risks in conducting contingency, steady state, and training missions. This report is delivered as part of an fiscal year 2015 PAF project titled "Support to Air Force Installation and Mission Support Center," cosponsored by Major General Theresa Carter, AFIMSC Commander, and Gilbert Montoya, Air Education and Training Command A4/7. The research was conducted within the Resource Management Program of PAF. This report should be of interest to AFIMSC personnel and major commands, component major commands and component numbered air forces, and others who will rely on AFIMSC operations for support.

This report was commissioned by AFIMSC and Air Education and Training Command Director of Logistics (AETC/A4) at a time when AFIMSC was being established to focus on managing the installation component of the ACS enterprise. The report highlights connections between AFIMSC (focused on installation support) and the Air Force Sustainment Center (focused on mission generation and sustainment support), both of whom must work together as key components of the ACS enterprise. The report also highlights a vision for ACS command and control (C2), an area that has long been the subject of RAND PAF research, and AFIMSC's role as part of the Air Force's ACS C2 capability. Since the commissioning of the study and completion of the draft report, both AFIMSC and the Air Force Sustainment Center have evolved in their roles related to the ACS enterprise and ACS C2. The authors recognize this evolution and suggest that readers view the recommendations giving due consideration to progress the Air Force has made since the study was conducted.

RAND Project AIR FORCE

RAND Project AIR FORCE (PAF), a division of the RAND Corporation, is the U.S. Air Force's federally funded research and development center for studies and analyses. PAF provides the Air Force with independent analyses of policy alternatives affecting the development, employment, combat readiness, and support of current and future air, space, and cyber forces. Research is conducted in four programs: Force Modernization and Employment; Manpower, Personnel, and Training; Resource Management; and Strategy and Doctrine. The research reported here was prepared under contract FA7014-06-C-0001.

Additional information about PAF is available on our website: http://www.rand.org/paf.

This report documents work originally shared with the U.S. Air Force on August 6, 2015. The draft report, issued on January 3, 2016, was scrutinized by formal peer reviewers and U.S. Air Force subject-matter experts.

Contents

Figures

Tables

Summary

The U.S. Air Force continues to operate in a challenging environment, defined by geopolitical uncertainty, multiple ongoing deployments to several regions, increasing threats to forward airbases, and fiscal constraints. The Air Force must shrewdly allocate its limited resources and, in line with the Air Force Future Operating Concept, do so in an agile manner. One of the ways the Air Force has sought to achieve better enterprise management of resources is to consolidate oversight under single organizations—e.g., some of the actions under the recent reorganization of Air Force Materiel Command.

More recently, acting on the direction of the Secretary of the Air Force and the Chief of Staff of the Air Force, the Air Force established the Air Force Installation and Mission Support Center (AFIMSC) with a focus on consolidating "management, oversight, and resources needed to support MAJCOM [major command] and installation commanders, as well as organize, train, and equip ACS [agile combat support] airmen."[1] As AFIMSC establishes operations consistent with guidance provided in Headquarters, U.S. Air Force, Program Action Directive (PAD) 14-04, it must make near-term resource allocation decisions across the installation and mission support (I&MS) enterprise, referred to in the PAD as *enterprise operations.*[2] Those decisions must be made using a transparent process that incorporates inputs from MAJCOMs and installation commanders who previously held responsibility for resource allocation decisions.

RAND Project AIR FORCE has worked with the Air Force for many years to develop ways to cope with these challenges, including analyses that provide a theoretical foundation for processes and policies guiding the operations of new ACS global management organizations.[3]

---

[1] Program Action Directive (PAD) 14-04, *Implementation of the Air Force Installation and Mission Support Center (AFIMSC)*, Washington, D.C.: Headquarters, U.S. Air Force, 2014, p. 1.

[2] PAD 14-04, 2014, p. 2.

[3] A series of RAND reports on command and control and enhanced ACS processes highlighted the value of organizations focused on engaging with warfighters to guide resource allocation in a manner that improves operational capability and effectiveness. Those analyses provided rationale for creating the Air Force Global Logistics Support Center, Air Force Sustainment Center, and AFIMSC. See James A. Leftwich, Robert S. Tripp, Amanda B. Geller, Patrick Mills, Tom LaTourrette, Charles Robert Roll Jr., Cauley von Hoffman, and David Johansen, *Supporting Expeditionary Aerospace Forces: An Operational Architecture for Combat Support Execution Planning and Control*, Santa Monica, Calif.: RAND Corporation, MR-1536-AF, 2002; Robert S. Tripp, Kristin F. Lynch, John G. Drew, and Robert G. DeFeo, *Improving Air Force Command and Control Through Enhanced Agile Combat Support Planning, Execution, Monitoring, and Control Processes*, Santa Monica, Calif.: RAND Corporation, MG-1070-AF, 2012; Kristin F. Lynch, John G. Drew, Robert S. Tripp, Daniel M. Romano, Jin Woo Yi, and Amy L. Maletic, *Implementation Actions for Improving Air Force Command and Control Through Enhanced Agile Combat Support Planning, Execution, Monitoring, and Control Processes*, Santa Monica, Calif.: RAND Corporation, RR-259-AF, 2014a; Kristin F. Lynch, John G. Drew, Robert S. Tripp, Daniel M. Romano, Jin Woo Yi, and Amy L. Maletic, *An Operational Architecture for Improving Air Force Command and Control Through Enhanced Agile Combat Support Planning, Execution, Monitoring, and Control Processes*, Santa Monica, Calif.: RAND Corporation, RR-261-AF, 2014b.

The commander of AFIMSC asked RAND to help shape a vision for analytical capabilities that will enhance resource requirements and allocation decisions across the planning, programming, budgeting, and execution system (PPBES) time horizon. This report responds to that request.

The status quo system that AFIMSC inherited for making resource allocation decisions as part of its enterprise operations responsibilities cannot simply be carried forward as is. In the past, each MAJCOM developed its own I&MS program objective memorandum (POM) inputs based on its own knowledge and expertise. But MAJCOMs (each of which has few bases and deep local knowledge) differ in their mission requirements, and installations differ in their support strategies (for example, the mix of military or civilian personnel and the mix of organic or contractor support), thus making the task of enterprise-wide resource allocation especially challenging.

Further, MAJCOMs differ in the processes they use to solicit requirements and allocate resources; there is no single method to replicate. Thus, those processes that depend so much on tacit local knowledge and direct communication (for example, between MAJCOM Headquarters and installation commanders) cannot be scaled and replicated easily at the enterprise level.

Thus, AFIMSC needs a method to allocate resources across missions and installations and to do so in a manner that is rational and inclusive. In this report, we describe such an approach.

Our approach has three elements:

- a lexicon of metrics to enable clear communication between AFIMSC and its customers
- transparent business rules to enable targeted resource allocations and illuminating performance measurement and reporting
- a reporting construct to aggregate performance information to make it digestible by its four-star customers.

The current system for identifying I&MS requirements is rarely output-focused. In other words, the consequences of different funding levels are not clear. To allocate I&MS funds rationally and transparently, AFIMSC needs, as much as possible, to tie funding levels to outcomes. One potential tool to help accomplish this is referred to as Common Output Level Standards (COLS), a system adopted by the joint community and the Air Force for defining standardized levels of installation and mission support.

Under COLS, standard levels of support have been identified for 34 I&MS activities, each with a subordinate set of measures that can be used to define four different levels of support, from full capability at level 1, down to greatly reduced capability at level 4, which is substandard but still complies with legal requirements and still supports the operational mission.[4] Some functions have established a further definition of each COLS level, with a breakout of the subprogram or activity measures that are considered to determine the COLS level. For example, for fire emergency services, the civil engineering community considers four key services with

---

[4] U.S. Air Force, *Air Force Common Output Level Standards*, PowerPoint presentation, Washington, D.C., December 2011.

associated measures and has established a detailed stratification: command and control, fire prevention, fire operations, and hazardous material operations. Each of these services then has a level of capability and a set of resources associated with each COLS level.

COLS-based metrics can help communicate customer requirements, but AFIMSC needs a set of business rules to rationally and transparently assess resource allocation options. Our proposed approach to developing these business rules has three elements:

- separate installation support from mission support activities (for the purposes of reporting performance and allocating resources)
- classify installations and mission activities to guide resource allocations
- develop business rules to rationally and transparently make allocations.

In our construct, installation support focuses on those functions that are critical to supporting an installation and its population, as well as on missions not directly tied to force projection. It is helpful to think of the installation as a municipality when considering which functions, activities, and resources fall into this category. We argue that, for the most part, these functions can be funded and supported at a single global level of service, such that an airman on any base would receive at least comparable, if not identical, levels of service. Examples include morale, welfare, and recreation; mortuary affairs, lodging, family support, contracting, and financial management, as well as select activities within security forces, civil engineering, and more.

Mission support includes activities for which AFIMSC would more often differentiate levels of support, driven by unique operational mission requirements. One example is the communications infrastructure for a command and control center or a satellite control center. Another example is the electrical power requirements of an aircraft maintenance hangar compared with a flying-simulator building. Examples of the activities in the mission support category include much of security forces; risk-focused civil engineering activities, such as explosive ordnance disposal and fire protection; and communications for flight line and air traffic control communication networks. For these, each installation or individual mission would argue for special requirements, why it requires a higher COLS level (that is, standard levels of support) than other bases or missions, and what level of resources correlate to each COLS level.

Classifying installations and missions goes hand in hand with developing the actual business rules. Each base has different requirements for installation and mission support, and depending on the types of operational mission they support, a degraded capability can have different effects on the Air Force's core missions and its ability to provide combat-ready forces to combatant commands. For example, a strategic nuclear base would presumably require higher than normal security for some areas, while a remotely piloted aircraft base or a satellite control installation would require robust and reliable communications to the installation itself and to key facilities. More specifically, one mission might require high base defense, while another might simply require more security for a particular facility or area (for example, military or civilian personnel, supplies, and equipment to perform preventive or emergency maintenance).

To help guide prioritization, we propose a taxonomy for classifying installations based on the missions they support and those missions' contribution to the Air Force's support to combatant commands. We also discuss examples of existing prioritization schemes that could inform the development of business rules to guide resource allocations.

The third and final element in our proposed system is a reporting construct. Senior leaders are often bombarded with metrics reports, frequently without a way to synthesize and digest all the information they are viewing. We therefore propose the development of a performance-monitoring construct patterned after an existing Air Force system that is as simple and transparent as possible.

Air Force Material Command uses the Weapon System Enterprise Review (WSER) to gain a comprehensive view of the status of weapon systems. The review looks at high-level areas, including performance execution, modernization, product support, and predictive health. Under each of these categories, there are subordinate measures, such as not mission capable for supply, not mission capable for maintenance, aircraft delivery quality, cost, schedule, health assessments of logistics, and leading health indicators for engines, as well as others. The status for each metric is reported and contributes to the overall "health" rating for each weapon system. The WSER includes a set of business rules that dictate how the overall rating should be adjusted if some number of subordinate metrics are degraded.[5]

We propose the creation of a WSER-like measurement system that considers installation support and mission support as enterprise capabilities composed of I&MS functions. The Installation Support Capability Rating (ISCR) and Mission Support Capability Rating (MSCR) would provide a comprehensive rating of the functions that are required to produce a full capability, as opposed to rating each individual functional area. This system could serve as a widely accepted lexicon for defining and measuring I&MS output capabilities.

The summary rating levels for an installation's ISCR and MSCR could be determined using a set of business rules that dictate when the overall rating is downgraded.[6] The criteria for different MSCR levels can be similar in nature to those for the ISCR but might also be tied to operational mission effect and risk associated with operational mission execution.

Establishing these business rules before implementing the proposed measurement system and then vetting them with stakeholders are critical steps in the implementation. This system should provide the primary communications mechanism and lexicon for the AFIMSC's engagement with its stakeholder. A degraded capability sends a strong signal to the AFIMSC that action is needed, similar to the way that WSER is viewed. As a part of the reporting system, each

---

[5] U.S. Air Force, *Weapon System Enterprise Review Business Rules*, Washington, D.C., June 2015.

[6] The Air Force uses a comprehensive set of business rules as part of the WSER. For each category measured in the WSER, the business rules specify under what conditions the category should be rated a certain color (green, yellow, red, or blue).

installation reporting a COLS level lower than acceptable would also be required to report a mitigation plan and associate costs with the mitigation plan.

The approach proposed here for near-term allocation decisions also supports long-term resource planning and allocation decisions. During the programming phase of the PPBES process, manpower and equipment standards and infrastructure funds for sustainment, restoration, and modernization would be aggregated into a total I&MS requirement. Individual bases would help translate their unique requirements into COLS levels so the funding requirements could be communicated in terms of local performance, and vice versa.

The POM build process could iterate with trade-offs between COLS levels and across installation categories until an acceptable budget request is developed for the entire I&MS enterprise and incorporated into the ACS core functional lead input. For example, the initial POM build might use as a baseline the assumption that all installations will be funded at an MSCR COLS level 1 and an ISCR COLS level 2. If, after the initial review of the proposed POM input, the requirement exceeds a reasonable level that the I&MS enterprise could expect to compete for within the Air Force Corporate Structure, a second iteration might include a modification where only certain installation types will be funded at an MSCR COLS level 1, while all other installation types are programmed at an MSCR COLS level 2.

Using this approach in the PPBE process provides several key benefits to the Air Force generally and the AFIMSC specifically. First, the process is inclusive and provides a mechanism for installations to make inputs. Second, it considers functional areas as capabilities composed of complementary, mutually exclusive, completely exhaustive components, in much the same way that Major Force Programs are viewed and compete within the PPBE process. Third, the approach serves as a mechanism to set expectations for the level of support that can be achieved at any given installation during the year of execution. Finally, after Congress approves the budget and funds are distributed, the approach establishes a baseline against which units can be measured during the year of execution. In addition, this process incorporates the enterprise view of resource allocation that the Air Force sought in establishing AFIMSC and assigning this new organization with that goal.

During the year of execution, reporting through the ISCR/MSCR system will reveal whether an installation's capability is below or above the level for which it is budgeted. If a unit is only funded to a COLS level 3 and it reports a COLS level 3, there is no need for rebalancing or redistributing resources by the AFIMSC. If a unit is funded for COLS level 2 but it is operating at a COLS level 3, its unfunded requirements would be candidates for additional funding. The same methodology could be used for making capability cuts when there are budget constraints. A deliberate and transparent decision could be made about where to allocate capability cuts and the cost savings that would be expected as a result of making those cuts.

Recommendations

To mitigate the challenges of the current security and fiscal environments and to capitalize on the transformative initiatives affecting the ACS community, we recommend that the Air Force

- implement an analytical framework for AFIMSC, as the supplier of I&MS resources, that provides meaningful trade-off information to both the customers of I&MS capabilities and an integrator responsible for adjudicating mismatches between demand and supply
- develop a lexicon and set of business rules to inform the decision tradespace of the operations community, in concert with and agreed on by the operational community
- implement an approach to an enterprise-wide measure system that is based on the WSER—which is in use at Air Force Materiel Command today—and includes a standardized capability-rating structure for installation and mission support capabilities
- employ a step-wise approach in implementing that system—this report lays out such an approach to implementing this framework.

Acknowledgments

Numerous people both within and outside the Air Force provided valuable assistance to and in support of our work. They are listed here with their rank and position as of the time of this research. We thank Major General Theresa Carter, Commander, Air Force Installation and Mission Support Center, and Gilbert Montoya, Deputy Chief of Staff, Logistics, Engineering, and Force Protection, Air Education and Training Command, for sponsoring this work. We also thank their staffs for their time and support during this research.

We thank Carolyn Gleason, our action officer (SAF/FMB). We also thank Lynn Eviston (AFMC/A8/9) for patiently explaining some of the background and original intent of the Air Force Installation and Mission Support Center. We thank Charlene Xander (AFMC/A1M) for providing information about Global Base Support.

At RAND, we thank Mahyar Amouzegar and Elvira Loredo for reviewing this work and helping to sharpen our analysis. We thank Jerry Sollinger for editing this report and helping to make it clear and concise. We thank Regina Sandberg for her help in shepherding the document through the process to publication.

Responsibility for the content of the document, analyses, and conclusions lies solely with the authors.

Abbreviations

AA	aircraft availability
ACS	agile combat support
AFFOR	Air Force forces
AFIMSC	Air Force Installation and Mission Support Center
AFSC	Air Force Sustainment Center
ALEX	Agile Logistics Evaluation eXperiment
C2	command and control
CC	commander
CCDR	combatant commander
CFL	core function lead
C-MAJCOM	component major command
C-NAF	component numbered air force
COCOM	combatant command
COLS	Common Output Level Standards
COMAFFOR	Commander, Air Force Forces
CS	combat support
DoD	Department of Defense
DRIVE	Distribution and Repair in Variable Environments
DRU	direct reporting unit
EXPRESS	Execution and Prioritization of Repair Support System
FYDP	Future Years Defense Plan
GIC	global integration center
HAF	Headquarters, Air Force
HAZMAT	hazardous material
I&MS	installation and mission support

ICP	inventory control point
IOC/FOC	initial operating capability/full operating capability
ISCR	Installation Support Capability Rating
JCS	Joint Chiefs of Staff
MAJCOM	major command
MICAP	mission-capable parts
MOE	measure of effectiveness
MSCR	Mission Support Capability Rating
MWR	morale, welfare, and recreation
NAF	numbered air force
OA	operational architecture
OPLAN	operational plan
Ops	operations
PAD	program action directive
POM	program objective memorandum
PPBES	planning, programming, budgeting, and execution system
PSU	primary subordinate unit
RSP	readiness spares package
SECDEF	Secretary of Defense
SPRS	Spares Priority Release Sequence
SRM	sustainment, restoration, and modernization
TOA	total obligation authority
UMMIPS	Uniform Materiel Movement and Issue Priority System
WSER	Weapon System Enterprise Review

1. Introduction

Background

Acting on the direction of the secretary and the chief of staff of the U.S. Air Force, the Air Force established the Air Force Installation and Mission Support Center (AFIMSC), with a focus on consolidating "management, oversight, and resources needed to support MAJCOM [major command] and installation commanders, as well as organize, train, and equip ACS [agile combat support] airmen."[7] Guidance for implementing AFIMSC is provided by Headquarters, U.S. Air Force (HAF), Program Action Directive (PAD) 14-04. The PAD outlines the following objectives:[8]

- Rebaseline HAF responsibilities, focusing MAJCOMs and direct reporting units (DRUs) on primary mission areas and focusing numbered air forces (NAFs) on operational mission execution.
- Reduce duplicative management oversight performed at all MAJCOMs and DRUs, more effectively and efficiently managing installation and mission support (I&MS) resources using an Air Force enterprise-wide view, and integrate advocacy for these critical capabilities.
- Consolidate I&MS resource planning and program objective memorandum (POM) development in AFIMSC as part of the broader ACS core function lead (CFL) portfolio.
- Align a diverse range of subordinate units as primary subordinate units (PSUs) under AFIMSC to centralize the management, oversight, and resources needed to support MAJCOM and installation commanders, as well as organize, train, and equip ACS airmen.[9]
- Deliver five overarching capability groupings: AFIMSC enterprise operations, installation management, protection services, installation operations, and expeditionary.

In addition to outlining the objectives of the reorganization, PAD 14-04 describes the structure of the organization—including such standard command staff functions as legal, personnel, safety, inspector general, and small business support and three directorates that will conduct large portions of AFIMSC operating functions. Those directorates are the Expeditionary Support Directorate, the Installation Support Directorate, and the Resources Directorate.[10] The PAD does not clearly delineate the roles and responsibilities of each directorate. It does,

---

[7] Program Action Directive (PAD) 14-04, *Implementation of the Air Force Installation and Mission Support Center (AFIMSC)*, Washington, D.C.: Headquarters, U.S. Air Force, December 2014, p. 1.

[8] PAD 14-04, 2014, p. 1.

[9] These PSUs include the Air Force Civil Engineering Center, the Air Force Security Forces Center, the Air Force Installation Contracting Agency, the Air Force Financial Services Center, the Financial Management Center of Excellence, and the Air Force Personnel Center Services Directorate.

[10] PAD 14-04, 2014, p. A-7.

however, reflect the fact that the first two directorates contain cross-functional staffs that, according to their titles, focus on ensuring support for both expeditionary operations and ongoing installation support. Fulfilling those responsibilities would require these directorates to understand mission and customer demands and manage resource allocations to meet those demands. The Resources Directorate appears to be the only activity with an assessment and analysis capability,[11] which suggests that the directorate is responsible for evaluating effectiveness, risk, and cost trade-offs among I&MS resources and capabilities across the worldwide enterprise using analytic tools and processes.[12]

Before the creation of AFIMSC, MAJCOMs provided installation management and oversight and tailored support strategies to their respective missions. Expeditionary requirements were levied on the MAJCOMs, as force providers, by the Commander of Air Force Forces (COMAFFOR) and in some cases by combatant commands (COCOMs). At its core, the creation of AFIMSC represents an enterprise-wide centralization of I&MS functions that were previously decentralized and managed by MAJCOMs, installation commanders, and associated functional communities (for example, civil engineering, security forces, communications).[13] The effect of this centralization on installation and MAJCOM commanders (MAJCOM/CCs) is explicitly recognized in the PAD. The guidance in the PAD highlights the need for AFIMSC to accomplish the following:

- provide responsive support to commanders using a transparent governance system and consistent, standardized business processes[14]
- develop standardized, collaborative, effective, and efficient governance processes to prioritize requirements for installations and provide feedback to respective functional leads, MAJCOMs, and the Air Force Corporate Structure[15]
- have governance processes that provide transparency and adequate avenues for supported commander input to key decisionmaking.[16]

Managing installation and mission support from an enterprise perspective is more complex than the scope of installations that any single MAJCOM formerly managed. The complexity of the enterprise is driven by differing mission demands, and resource allocation decisions will routinely involve trade-offs among effectiveness, efficiency, and risk across those demands. This decisionmaking should be supported by an analytical framework that considers multiple alternatives. New processes that provide insights for support alternatives might be required to

---

[11] PAD 14-04, 2014, p. A-7.

[12] Discussions with key participants in the planning for AFIMSC, as well as with participants involved in standing up AFIMSC, confirmed that the Resources Directorate is intended to provide analysis supporting trade-off decisions.

[13] PAD 14-04, 2014, p. A-2.

[14] PAD 14-04, 2014, p. 3.

[15] PAD 14-04, 2014, p. 9.

[16] PAD 14-04, 2014, p. 9.

shift resources across the enterprise and to ensure successful accomplishment of the operational missions required by COCOMs and COMAFFORs. These trade-offs should be informed by a set of metrics that capture the missions of the AFIMSC customer base, such as operational effects, home-station training, deploy-to-dwell ratios, and population support levels of services.

Transitioning the management of I&MS resources from MAJCOMs to a centralized organization—in this case, AFIMSC—could present concerns for the MAJCOMs. How can a centralized organization understand the unique requirements of their MAJCOM forces? How will resource allocation decisions be made when several MAJCOMs request allocation of scarce resources? Analysis can help address stakeholder concerns through open and clear communication about how requirements will be determined, how assumptions and options for addressing requirements will be vetted, and how metrics will be used in trade-off decisions. The analysis process needs to be transparent to ensure that stakeholders focus on the substance of how to address requirements; how to state costs, risks, and effectiveness of alternative means of satisfying the requirements; and, if requirements cannot be met, how they can be adjudicated or mitigated equitably. To provide this analysis, AFIMSC requires a solid analytical framework that is structured and theoretically sound.

Purpose

For many years, RAND Project AIR FORCE has helped the Air Force develop a conceptual model and associated analytical framework for managing global ACS functions and better integrating them with operational planning and execution. The Air Force has asked RAND to help shape a vision for analytical capabilities that will enhance resource requirements and allocation decisions across the planning, programming, budgeting, and execution system (PPBES) time horizon.

To that end, this report distills insights from prior RAND research that can be applied to AFIMSC. The report describes a conceptual model that clarifies the relationships between customers (i.e., demanders) and suppliers of combat support resources and the need for analysis to facilitate decisionmaking among these stakeholders. It also outlines an analytical framework for conducting such analyses and how it can be applied to AFIMSC. Finally, the report proposes a new lexicon and measurement structure for IM&S capability that will enable AFIMSC to work closely and communicate with the customers of I&MS resources. The appendixes describe prior RAND research that illustrates the types of analyses that are enabled by these theoretical and analytical frameworks.

How This Report Is Organized

This report has five chapters and four appendixes. Chapter Two describes the origin of the theoretical underpinning for AFIMSC to operate as an enterprise manager and presents a conceptual model for balancing supply and demand, including an approach for prioritizing decisions when the latter exceeds what the former can supply. Chapter Three describes the proposed analytical framework for AFIMSC in its role overseeing the I&MS enterprise. Chapter

3

Four addresses and proposes an approach for implementing mechanisms that are necessary for the analytical framework to function properly. Chapter Five presents conclusions and recommendations and describes what actions should be taken next.

This report has four appendixes. Appendix A summarizes past RAND research on ACS command and control (C2). This provides more detail behind the history presented in Chapter Two. Appendix B summarizes past, unpublished research that proposes and demonstrates methods for quantifying the effects of home-station ACS shortfalls. This can be used to inform AFIMSC efforts to balance expeditionary versus home-station needs. Appendix C presents an illustrative alignment of common output-level standards (COLS) to Installation Support and Mission Support Capability Ratings, categorizations we present in Chapter Four. Appendix D presents an illustrative mapping of program element categories to Air Force COLS. This supports analysis presented in Chapter Four.

2. An Operational Framework for AFIMSC as an Enterprise Manager

This chapter describes the conceptual model for AFIMSC to use in balancing demand and supply as an enterprise manager. The chapter begins by highlighting a model developed by RAND several years ago and follows that discussion with one that applies the conceptual model to AFIMSC. Appendix A provides additional information on prior RAND research that is referenced in developing the model for AFIMSC.

An Operational Architecture for Enhancing ACS Planning, Execution, Monitor, and Control Processes

Shortly after the completion of Operation Noble Anvil,[17] and in the midst of the Air Force's transition to a more expeditionary force employment concept, RAND was asked by AF/A4 (Deputy Chief of Staff for Logistics, Engineering and Force Protection, Headquarters U.S. Air Force) to develop a concept for ACS execution, planning, monitoring, and control. The objectives were to develop an operational architecture for enhancing ACS processes within the Air Force C2 system that addressed the shortcomings exposed by Operation Noble Anvil and would enable the ACS community to meet the evolving demands of being an expeditionary Air Force.

RAND completed the first body of research in 2002 and provided an in-depth operational architecture (OA) consistent with the Department of Defense (DoD) architectural framework construct.[18] The OA mapped ACS processes across the strategic, operational, and tactical levels, from the Secretary of Defense (SECDEF) and the Joint Chiefs of Staff (JCS) down to the unit level and across phases of an operation from planning and programming to reconstitution. A key characteristic of the OA was that it not only mapped the ACS processes but also addressed the integration of those processes within the PPBES and the corporate Air Force operational C2 enterprise.

The OA highlighted a number of key points about ACS planning, execution, monitoring, and control within the context of the Air Force's larger C2 processes. First, the OA showed the global, interactive, and multiechelon nature of ACS planning, execution, monitoring, and control within the broader Air Force C2 enterprise. The OA communicated the roles of Air Staff,

---

[17] The U.S. portion of Operation Allied Force in Serbia during 1999 was code-named Operation Noble Anvil.

[18] James A. Leftwich, Robert S. Tripp, Amanda B. Geller, Patrick Mills, Tom LaTourrette, Charles Robert Roll, Jr., Cauley von Hoffman, and David Johansen, *Supporting Expeditionary Aerospace Forces: An Operational Architecture for Combat Support Execution Planning and Control*, Santa Monica, Calif.: RAND Corporation, MR-1536-AF, 2002.

MAJCOMs, Air Force forces (AFFOR) staffs, NAFs, and installations in commanding and controlling the ACS enterprise to meet warfighter needs.

Second, the OA demonstrated the iterative and interactive role of ACS planning and execution with operational planning and execution. Finally, the OA introduced a baseline process (shown in Figure 2.1) that included establishing a plan, developing measures of effectiveness (MOEs) related to operational performance, assessing actual performance against planned performance, and adjusting ACS resources as needed to ensure that MOEs remained within acceptable thresholds. That baseline process was repeated throughout the OA across the echelons and phases of operation.

Figure 2.1. Closed-Loop Planning Process Used in Enhanced ACS Process Across the Phases of an Operation

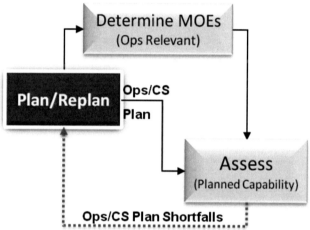

NOTE: Ops/CS = operations/combat support

Many of the process and doctrine recommendations formed the baseline for the analytical framework that we propose for AFIMSC. The process recommendations are depicted throughout the OA and incorporate key points, such as integration with operations and the inclusion of closed-loop feedback.

The creation of AFIMSC provides the Air Force with an organization capable of fulfilling the ACS planning, execution, and control functions recommended in the original and subsequent RAND research discussed in more detail in Appendix A. The recommendations from the extended body of research form the baseline for the analytical processes and framework proposed for AFIMSC.

A Conceptual Model for Managing an Enterprise in a Resource-Constrained Environment

The challenge before AFIMSC is to provide a range of I&MS capabilities to a diverse customer base within a given budget ceiling. Prior RAND research outlined a method to allocate scarce resources from a strategic planning perspective.[19] The research outlined a model that recognizes the roles for customers (i.e., demanders) of resources, suppliers of resources, and an integrator that makes choices regarding which demands to satisfy when resources are constrained.

Figure 2.2 illustrates the resource-constrained framework at the highest operational levels within DoD. Here, COCOMs are the customers (demand side) and develop requirements and priorities (for military forces, for example). The military services act as force providers, or suppliers of capabilities. The SECDEF acts as an integrator to resolve disputes or imbalances, if and when the demand exceeds the supply. This arrangement applies in a policymaking or deliberate planning environment—for example, the apportionment of forces to operational plans (OPLANs), as well as during execution (for example, if multiple COCOMs request the deployment of a scarce unit type or resource).

The model applies at lower levels of command as well, whether at the wing level, AFFOR level, or MAJCOM level. Force apportionment is a well-known application of this arrangement. Another example is the allocation of scarce precision munitions. This resource allocation model has been well documented by RAND in prior analyses and vetted by both operational and combat support communities within the Air Force.[20]

---

[19] See Leslie Lewis, James A. Coggin, and C. Robert Roll, *The United States Special Operations Command Resource Management Process: An Application of the Strategy-to-Tasks Framework*, Santa Monica, Calif.: RAND Corporation, MR-445-A/SOCOM, 1994. The model referenced here for allocating scarce resources across competing demands, while first introduced in the study by Lewis et al., has been a cornerstone of more-recent studies by RAND addressing ACS C2.

[20] For discussions on the deficiencies identified through these research efforts, see Leftwich et al., 2002; Kristin F. Lynch, John G. Drew, Robert S. Tripp, and Charles Robert Roll, Jr., *Supporting Air and Space Expeditionary Forces: Lessons from Operation Iraqi Freedom*, Santa Monica, Calif.: RAND Corporation, MG-193-AF, 2005; Robert S. Tripp, Kristin F. Lynch, John G. Drew, and Robert G. DeFeo, *Improving Air Force Command and Control Through Enhanced Agile Combat Support Planning, Execution, Monitoring, and Control Processes*, Santa Monica, Calif.: RAND Corporation, MG-1070-AF, 2012; and Lynch, Kristin F., John G. Drew, Robert S. Tripp, Daniel M. Romano, Jin Woo Yi, and Amy L. Maletic, *Implementation Actions for Improving Air Force Command and Control Through Enhanced Agile Combat Support Planning, Execution, Monitoring, and Control Processes*, Santa Monica, Calif.: RAND Corporation, RR-259-AF, 2014a; Kristin F. Lynch, John G. Drew, Robert S. Tripp, Daniel M. Romano, Jin Woo Yi, and Amy L. Maletic, *An Operational Architecture for Improving Air Force Command and Control Through Enhanced Agile Combat Support Planning, Execution, Monitoring, and Control Processes*, Santa Monica, Calif.: RAND Corporation, RR-261-AF, 2014b.

Figure 2.2. Conceptual Model for C2 in a Resource-Constrained Environment

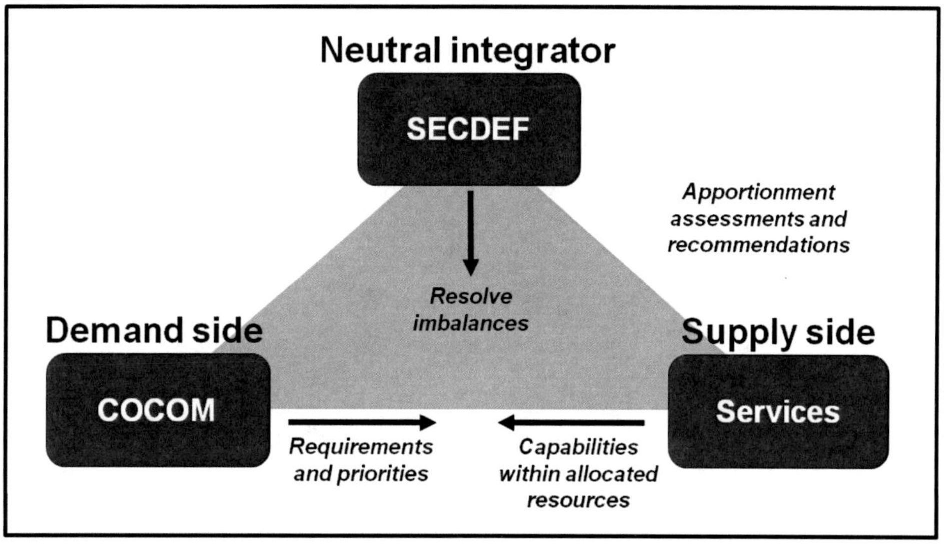

SOURCE: Tripp, Lynch, Drew, and DeFeo, 2012, p. 33.

The model recognizes the natural tension that exists between organizational elements responsible for delivering a product (that is, operational effects) and those charged with providing the necessary resources to deliver those products effectively and efficiently.[21] Given that natural tension, the framework introduces an integrator to adjudicate competition between supply and demand. Two core principles govern the model.

- **Separation of supply-side and demand-side decisions**. Supply- and demand-side decisions should be made separately and iteratively. Following this principle, the demand side specifies operational requirements and priorities for combat support resources, and the supply side decides how to satisfy those needs. The demand side does not instruct the supply side on how to provide required resources but specifies when capabilities are needed (to the extent that they are known). The supply side determines how the capabilities are to be met to efficiently satisfy the operational requirements within the time frame needed.
- **Independence of the integrator**. The integrator should be independent of both supply-side and demand-side organizations. If the integrator is too close to the supply side, then actions may lean toward efficiency at the expense of operational effectiveness. If, on the other hand, the integrator is too close to the demand side, then current operations may be effective without giving enough consideration to the efficient use of resources.

This conceptual model can help inform the roles and responsibilities of AFIMSC operational directorates and the capabilities required in those directorates.

---

[21] Tripp, Lynch, Drew, and DeFeo (2012) discussed the application of this framework in detail and included specific examples.

Conceptual Model Applied to AFIMSC

Applying the aforementioned model to the AFIMSC implementation provides a view of the challenges the organization faces and suggests the type of analysis AFIMSC might require to respond to those challenges. Figure 2.3 and the sections that follow illustrate that the model could be applied to AFIMSC.

Figure 2.3. Conceptual Model Applied to AFIMSC

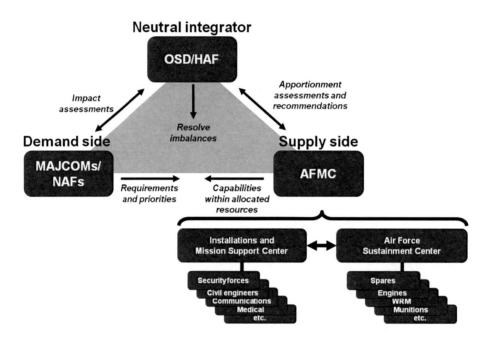

SOURCE: This figure is drawn from several figures first highlighted in Robert S. Tripp, John G. Drew, and Kristin F. Lynch, *A Conceptual Framework for More Effectively Integrating Combat Support Capabilities and Constraints into Contingency Planning and Execution,* RAND Corporation, RR-1025-AF, 2015.

NOTE: OSD = Office of the Secretary of Defense; AFMC = Air Force Materiel Command; WRM = war reserve material.

The Demand Side

The AFIMSC customer base is broad and diverse. It includes MAJCOMs responsible for organizing, training, and equipping forces that will deploy in support of COCOMs; U.S.-based installations that provide global reach and power projection from home stations; component MAJCOMs (C-MAJCOMs) and component NAFs (C-NAFs) responsible for supporting theater COCOMs that develop plans for establishing installations during contingencies; and installations whose primary missions focus on acquisition, research and development, education, and others. Each customer has I&MS requirements that AFIMSC must meet. AFIMSC has detachments at MAJCOMs that can help ensure that I&MS demands are transmitted to AFIMSC and are clearly understood.

One of the complexities of the demand side is that I&MS resource requirements are driven by various external factors. In some instances, resource requirements (e.g., manpower and equipment) are derived from peacetime home-station use rates; in others, resource levels are established by contingency requirements. Other factors, such as the type of installation and the operating environment, can also drive resource demands. Some installations perform their contingency roles from their home stations. Other installations support missions that are focused on research and development or education, which are less affected by contingency operations. Each installation and MAJCOM has tacit knowledge of these complexities that AFIMSC will have to deal with. The measurement system and reporting structure suggested later in this report are intended to provide installation commanders and MAJCOMs a mechanism for communicating those complexities and the impact of I&MS resourcing on mission accomplishment.

As customers are provided I&MS capabilities, they must develop assessments to quantify and articulate how the capabilities being provided are supporting their missions and activities, as well as the nature and impact of any shortfalls.

The Supply Side

As discussed in Chapter One, AFIMSC's role is as the supplier of installation and mission support capability. As depicted in Figure 2.3, AFIMSC must allocate resources to meet the demands of the COCOMs and COMAFFORs in wartime, as well as the demands of the MAJCOMs and installations during peacetime and wartime. It must interpret the capabilities desired by the demand side, translate those demands into resources requirements that are organized within functional stovepipes, and manage those resources across the enterprise. AFIMSC must make trade-offs among competing demands and manage resource constraints resulting from those competing demands. AFIMSC must continue to adjust resource levels and make short-term and long-term investment decisions but do so, and communicate those decisions, in terms of the I&MS capabilities they provide and the operational effects that can be achieved. This is the same role that the Air Force Sustainment Center (AFSC) must perform with respect to managing the enterprise of assets that directly tie to sortie production (e.g., fuels support, maintenance, supply).

The Integrator

The third leg of the stool (top box in Figure 2.2) is the role of the integrator. For the I&MS enterprise, this role will sometimes be filled by HAF and sometimes by the Office of the Secretary of Defense. This role entails providing guidance for AFIMSC to develop trade-offs and capability options and directly arbitrating competing MAJCOM and COCOM demands, whether in POM development or execution. One example of proactive guidance in other systems is the Uniform Materiel Movement and Issue Priority System (UMMIPS), which prioritizes categories of movement demands. Another is the Spares Priority Release Sequence (SPRS) applied in

Execution and Prioritization of Repair Support System (EXPRESS) for the provision of spare parts.[22] These are both high-level prioritization schemes, produced and adjudicated separately from the supplier of capabilities, which the supplier simply applies. These schemes are open, transparent, and well understood by all stakeholders and thus enable buy-in.

Implications of the Conceptual Model on AFIMSC

The creation of AFIMSC provided an opportunity to realize the potential value of the three roles we have outlined. Customers of I&MS capabilities will need to further develop their abilities to quantify and articulate their mission requirements to outsiders who lack the tacit knowledge they have. Some of that will take the form of educating AFIMSC personnel; some may involve the work of translating local mission needs into a broader lexicon to enable those needs to be weighed against the full set of Air Force mission needs. With both, there is significant work to be done and dramatic shifts in thinking required of ACS personnel.

Likewise, the guidance that AFIMSC will need to develop and weigh options does not yet exist. In the past, HAF allocated each MAJCOM a portion of the total Air Force total obligation authority (TOA), and the MAJCOMs in turn allocated to their installations as needed. In the future, some set of guidance and decisionmaking will be required to set up a scheme that is transparent and agreeable to stakeholders and to shape AFIMSC's analysis. And adjudication and arbitration responsibilities will need to be assigned and clarified to address conflicts in execution.

Finally, AFIMSC itself is evolving and exploring the analytic capabilities it will need. The diversity of demands it must support will generate questions that AFIMSC must be prepared to answer. These include, for example, the following:

- How will AFIMSC structure options for "supplying" MAJCOM and NAF "demands" for I&MS resources, including unique area of responsibility and installation requirements?
- What metrics should be used to evaluate I&MS resource options for both home-station and contingency support?
- What process should be used to structure analyses and vet effectiveness, risks, and costs of alternative resource allocations?
- How transparent will I&MS resource options be (e.g., underlying assumptions, data accuracy, model validity)?
- How can AFIMSC be responsive to changing customer needs?

Perhaps in anticipation of these questions, PAD 14-04 repeatedly levies the requirement for AFIMSC to establish governance processes, metrics, and implementation activities and to operate transparently in dealing with customers who previously were responsible for the

---

[22] EXPRESS is the system the Air Force developed to apply an approach called Distribution and Repair in Variable Environments (DRIVE). See John Abell, Louis W. Miller, Curtis E. Neumann, and Judith E. Payne, *DRIVE (Distribution and Repair in Variable Environments): Enhancing the Responsiveness of Depot Repair*, Santa Monica, Calif.: RAND Corporation, R-3888-AF, 1992.

management of the assets.[23] The following chapters of this report provide a framework from which AFIMSC can operate and answer these questions in a rational and transparent method.

---

[23] PAD 14-04, 2014.

3. Analytical Framework to Support AFIMSC Decisions

This chapter introduces an analytical framework to help AFIMSC shape a vision for analytical capabilities that will enhance resource requirements and allocation decisions across the PPBES time horizon.

We attempted to clarify the scope of questions AFIMSC will need to address to focus better on specific types of needed analytic capabilities and examples of the types of analyses (highlighted in the appendixes). Table 3.1 shows those questions.

Table 3.1. Questions That Inform Needed AFIMSC Analytic Capabilities

Topic	Execution	FYDP and Beyond
Peacetime steady state support	How to allocate sustainment, restoration, and modernization and other monies in *year of execution* to satisfy diverse installation requirements?	How to *shape strategies* for supporting installations that are cost-effective and balance competing demands?
Contingency expeditionary support	How to allocate resources in *contingency* to balance competing COCOM, C-MAJCOM, and C-NAF demands?	How to *shape* installation and mission support *posture* to meet contingency demands?

NOTE: FYDP = Future Years Defense Plan.

Providing reliable support to installations in the year of execution will require understanding unique mission demands and unique MAJCOM and installation support strategies (for example, military versus civilian versus contractor support), as well as developing tools that can tie strategies together to assess the impacts of different resource allocations (for example, a 10,000-person cut to military end strength).

To support installations through the FYDP and beyond, strategies are needed to shape the kinds of support that will be most cost-effective and balance home-station and expeditionary demands. Analysis developing home-station metrics (summarized in Appendix B) describes methods and metrics to assess the impact of deployments on home-station operations in a way that has the potential to inform installation support strategies. This includes quantifying and articulating "break the base" thresholds and specifying home-station installation performance and sustainability guidelines. Another stream of research, transaction cost analysis, synthesized business literature pertaining to insourcing and outsourcing decisions.[24] Shaping installation support over the long term will require trade-offs among many options.

To support contingency operations during execution, AFIMSC will need to assess the sufficiency of resources to support contingency demands and allocate those resources. RAND's

---

[24] John G. Drew, Ronald G. McGarvey, and Peter Buryk, *Enabling Early Sustainment Decisions: Application to F-35 Depot-Level Maintenance*, Santa Monica, Calif.: RAND Corporation, RR-397-AF, 2013.

C2 work outlines a concept and specific recommendation for better integrating combat support considerations during crisis action planning and execution.[25] This work tightly links ACS planning with operations planning and highlights the need to evaluate those linkages through experimentation.[26] In 2011, the Air Force demonstrated this concept in an Agile Logistics Evaluation eXperiment (ALEX) organized by the Air Force C2 Integration Center. The Operational Support Facility at Langley Air Force Base's Ryan Center hosted the AFFOR reach-back cell and provided ACS resource assessments for use in planning and supporting contingency operations in the European, Pacific, and Korean areas of responsibilities.

Finally, to prepare better for these contingency operations in the long term, AFIMSC can inform a balanced portfolio of ACS capabilities. RAND research (summarized in Appendix A) assessed the Air Force's total force ACS manpower mix and proposed recommendations to better prepare the Air Force for future scenarios.[27] This analysis had four main objectives: (1) assess the supply of ACS forces; (2) assess future expeditionary demands; (3) assess home-station requirements; and (4) assess policy options for better shaping the mix of ACS forces to meet future demands, while accounting for their ongoing home-station missions.

The remaining sections in this chapter focus primarily on installation support responsibilities and presents RAND's recommended analytic approach to near-term resource allocation decisions for installation support. Chapter Four presents our recommendations for how to institutionalize that approach.

Guiding Principles

The proposed analytical framework is built on several guiding principles. From a strategic perspective, the framework should do the following:

- **Relate I&MS resources to operational capabilities.** This principle is fundamental to the theoretical model that considers operations and combat support trade-offs when resources are limited. Effectiveness and efficiency analyses must be linked to operational capability, and I&MS shortfalls must be related to their effect on operations.
- **Integrate expeditionary and home-station requirements.** The demand for I&MS resources to meet both expeditionary and home-station requirements stands at the center of the planning and programming challenge for AFIMSC and must be addressed by the analytical framework.
- **Take an enterprise approach to shaping I&MS capabilities.** The establishment of AFIMSC provides an opportunity to consider I&MS resources from an enterprise-wide

---

[25] See Tripp, Lynch, Drew, and DeFeo, 2012.

[26] See Leftwich et al., 2002; Tripp, Lynch, Drew, and DeFeo, 2012; and Lynch, Drew, Tripp, Romano, et al., 2014a.

[27] Patrick Mills, John G. Drew, John A. Ausink, Daniel M. Romano, and Rachel Costello, *Balancing Agile Combat Support Manpower to Better Meet the Future Security Environment*, Santa Monica, Calif.: RAND Corporation, RR-337-AF, 2014.

perspective. To achieve the efficiencies and effectiveness intended by creating AFIMSC, the analytical framework must support the integration of I&MS resource management.

- ***Focus on long-range operational objectives.*** To function across the PPBES horizon, there must be a long-term view. Force structure shaping and reshaping in the context of future security environments are activities that often take years to achieve, and those decisions often require midcourse pivots as competing visions of the future security environment evolve.

An AFIMSC analytic framework guided by these principles facilitates decisionmaking in a transparent, timely, and theoretically sound manner. The framework enables operational and logistics stakeholders to make trade-off decisions with a common view of I&MS resource status, resource allocation options, and the effect of decisions on operations.

These guiding principles, together with the conceptual model highlighted in Chapter Two, lead to a structured analytical process that can be used by AFIMSC. Figure 3.1 highlights the core process around which the analytical framework can be established.

Figure 3.1. Core Process That Can Guide the Operation of AFIMSC's Analytical Framework

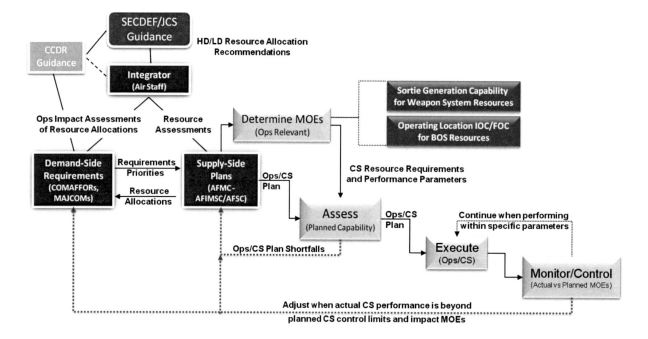

SOURCE: Tripp, Lynch, Drew, and DeFeo, 2012, p. 41.

NOTE: CCDR = combatant commander; HD/LD = high demand/low density; AFMC = Air Force Materiel Command; CS = combat support; IOC/FOC = initial operating capability/full operating capability; BOS = base operating support.

In Figure 3.1, the three blue boxes mirror the three boxes in Figure 2.2—supplier, demander, and integrator. These receive guidance from the purple boxes and incorporate it into their processes. Demanders (i.e., customers) determine and communicate requirements; AFIMSC develops and assesses supply-side (that is, resource allocation) plans and allocates resources. The resource allocation process is not simply unidirectional but includes high-level vetting and feedback from customers. Once a supply-side plan is made, that plan is executed, and both customers and AFIMSC monitor performance (the red dotted feedback arrows in the figure).

An important element of this process is developing appropriate MOEs. In responding to customer needs and expectations, the output of the analysis must be communicated using metrics that are meaningful to the customers. It is not sufficient to describe I&MS capabilities in terms of numbers of people or pieces of equipment. Instead, the metrics must inform discussion that occurs between installation commanders and COCOMs as customers and AFIMSC as suppliers of I&MS capabilities.

COCOMs' interest will center on such metrics as the number of forward operating locations that can be supported, the time required to establish initial and full operating capabilities at those forward locations, and the scalability of I&MS capability in response to surge demands. Home-station installation commanders will likely be interested in metrics focused on efficiency and costs, scalable levels of support when forces deploy from the base in response to contingencies, rotational stress, and the ability to maintain workforce expertise during contingencies or the time to rejuvenate it when degraded, among other metrics. Those types of operationally focused metrics will enable the critical trade-off decisions that must be made by consumers of I&MS resources, and translating the portfolio of I&MS resources into such metrics will be a key task of those analyses.

Within this analytical framework, AFIMSC is the supplier of I&MS resources and capabilities and will develop supply options using analytic processes. Installation commanders, MAJCOMs (component and supporting), and C-NAFs develop I&MS requirements in their role as consumers of installation and mission support capability. In this construct, there will be a natural tension between organizational elements responsible for delivering a product (i.e., operational capability) and those charged with providing the resources necessary to deliver those products effectively and efficiently.[28] Consistent with the conceptual model described in Chapter Two, HAF operates as the integrator and will adjudicate allocation decisions when the demand for resources exceeds supply and when the supply side and demand side are unable to reach an acceptable decision.

The Challenge

Three of the key elements in the resource allocation cycle as depicted in Figure 3.1 are: develop requirements, assess resource allocation options, and monitor performance. The Air Force's current process to determine requirements for I&MS resources and activities is mostly bottom-up. Units are authorized certain levels of manpower and equipment based on mission

---

[28] Tripp, Lynch, Drew, and DeFeo (2012) discussed the application of this framework in detail and included specific examples.

requirements and unique circumstances, according to published standards, allowances, and cost factors. For example, food service personnel are computed based on a formula driven by the number of enlisted personnel on base, plus an allowance for overhead (that is, a fixed-cost element), plus additional allowances, such as a flight kitchen.[29] Similar allowance standards exist for equipment.

Likewise, infrastructure sustainment, restoration, and modernization (SRM) funding requirements are computed with an enterprise-level model that applies sustainment cost planning factors to each infrastructure asset, classified by type. Those asset-level cost estimates get rolled up to the enterprise level to inform the POM. At each level of aggregation, priorities are generally transmitted, with an emphasis on local knowledge and preferences, or are automatically generated based on unit characteristics.

Resource allocation is then top-down, going from Congress, to the HAF, to the MAJCOMs. When MAJCOMs or functional centers get less than their fully stated requirement, they levy cuts within their MAJCOMs or functional domains, and they do their best to allocate funds based on some combination of mission impacts and a sense of equitability (including priority information that was transmitted during the requirements process).

Finally, performance is monitored in the course of mission accomplishment; problems are identified, with smaller ones being rectified to the degree possible with immediately available actions and resources and larger ones being kicked up to the next-highest-level resource allocation process. All of these processes happen with a mix of objective and subjective information and an ongoing dialogue among stakeholders about resource needs, shortfalls, and mitigation actions.

But AFIMSC cannot simply take over the status quo. Whereas the prior process allowed for optimization within a MAJCOM, AFIMSC must now balance across the entire enterprise characterized by a mix of missions and associated I&MS requirements. MAJCOMs (each of which has few bases and deep local knowledge) differ in their mission requirements, and installations differ in their support strategies (for example, military or civilian personnel, organic or contractor support). Further, MAJCOMs differ in the processes they use to solicit requirements and allocate resources; there is no single method to replicate. Thus, those processes that depend so much on tacit local knowledge and direct communication (for example, between MAJCOM headquarters and installation commanders) cannot be scaled and replicated easily at the enterprise level.[30] In a relatively stable and consistent funding environment, there already exists the risk (and worry on the part of some customers) that resource allocation decisions will

---

[29] U.S. Air Force, *Manpower Standard: Combat Support Flight*, AFMS 45XA, Washington, D.C., March 30, 2005.

[30] One principle in the field of organizational design is that decisions should be made by those who have the necessary information, something AFIMSC does not yet have. See Michael C. Jensen and William H. Meckling, "Specific and General Knowledge and Organizational Structure," in Lars Werin and Hans Wijkander, eds., *Contract Economics*, Oxford: Basil Blackwell, 1992.

not appropriately take these factors into account and will inevitably compromise mission accomplishment. If funding becomes unstable or unpredictable, this risk increases.[31]

Thus, AFIMSC needs a method to allocate resources across missions and installations and do so in a manner that is rational and inclusive. Its purpose is to relate the full complement of I&MS resources to output-focused performance metrics, provide a framework for measuring the impact of investments and cuts on I&MS performance, and construct an overarching taxonomy for classifying installations and missions that will assist in prioritizing investments and cuts across Air Force installations.

Fully implementing these processes will require the development of a lexicon of metrics to enable clear communication between AFIMSC and its customers and of transparent business rules to enable targeted resource allocations and illuminating performance measurement and reporting. As we survey the landscape of current Air Force concepts and capabilities, we find that our approach cannot be immediately and seamlessly implemented today. However, as we describe a vision for how AFIMSC can provide I&MS capabilities in a transparent, mission-oriented, and enterprise-focused way, we enumerate the tasks that need to be accomplished to realize that vision and describe steps along the way to help transition AFIMSC to its ultimate objectives.

Process View of AFIMSC Analytical Framework

Figure 3.2 provides a detailed view of an analytical process proposed for AFIMSC to manage the enterprise. This process is based on the conceptual model discussed in Figure 2.1 and incorporates activities that support the guiding principles. A similar process was developed and used by RAND to assess an enterprise view of ACS manpower-rebalancing options to meet future operational scenarios.[32] The research behind this is discussed in more detail in Appendix B.

Analytic Inputs

The framework starts with analytic inputs, including demand side and supply side. Those inputs come from a variety of sources and must be fully understood and vetted with stakeholders.

The first major input is the demand side, the requirements for I&MS capabilities from each of AFIMSC's customers. These inputs may be OPLANs or other planning scenarios, some other statement of contingency demands, or home-station installation requirements of various forms. (In the figure, we have included guidance as a demand-side input, in the form of planning scenarios.) Extracting requirements from operational planning guidance and scenarios is not necessarily straightforward. In this process, customers must articulate and quantify their

---

[31] Resource cuts are often disproportionately allocated to support functions, such as those in AFIMSC's purview, so there is some reason to worry.

[32] Mills et al., 2014.

requirements in such a way that I&MS suppliers can understand and quantify those needs and weigh them explicitly against other demands. For example, operations planners may state demands as the need to have three forward operating locations capable of supporting 24-hour-a-day combat sortie generation for two fighter squadrons that can be available within seven days. Requirements stated in this manner provide I&MS planners the insights they need to size civil engineering, security forces, services, and other I&MS functions to meet the needs.

Figure 3.2. Process View of the AFIMSC Analytical Framework

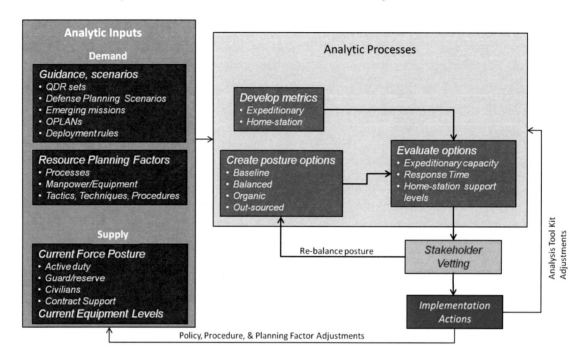

NOTE: QDR = Quadrennial Defense Review.

Until the formation of AFIMSC, there was no single clearinghouse to integrate and deconflict the range of expeditionary demands. In execution, the Aerospace Expeditionary Force Center within the Air Force Personnel Center was able to help identify available ACS manpower forces to support emerging and ongoing operations. But during planning and programming, several entities played a role. Career fields were able to look across a range of demands for their skill sets. The ACS CFL was able to advocate for options but had no directive authority. But, ultimately, resource allocation decisions occurred within the POM panels, and ACS manpower, as an example, resides within multiple panels.

On the installation side, customers must also articulate their requirements for the range of missions—including pilot training; mobility support to ongoing operations; research, development, test, and evaluation; and supply chain management—and also quality-of-life support to military members, their families, retirees, and so on.

The second input to the analysis is the supply side, which is built from the current I&MS resource and capability baseline. That baseline should consider the entire complement of resources—including active duty, Air National Guard, and Air Force Reserve Command personnel (both military and civil service); equipment; and resources committed to the enterprise through outsourcing (contract support) or arrangements with host nations or coalition partners. The current postures must be considered at the unit level and in aggregate, as well as within and across I&MS functions.

The third and final input in Figure 3.2 consists of guidance, essentially policy and instructions to AFIMSC, to integrate the demand and supply according to agreed-on inputs from stakeholders. This includes specific resource planning factors, processes, and standards of support for I&MS functions, as well as tactics, techniques, and procedures that drive the requirements for installation support. For example, some I&MS functional areas derive their resource requirements from contingency demands, while others derive their requirements from planned level of support for ongoing home-station peacetime operations. Some functional areas have varied levels of contingency response options that the COCOM must choose from based on acceptable levels of risk. Those planning factors, along with the guidance and scenarios, will likely provide a complete view of the demand for ACS capabilities.

Establishing an Analytical Capability for AFIMSC

AFIMSC can establish a capability that supports both the near-term and longer-term analytical needs that span the PPBES horizon and, in doing so, meet the requirement for transparency, governance, and stakeholder communications.

The framework described here can serve as a baseline to foster clearer guidance about future objectives through stronger collaboration with MAJCOMs, NAFs, and higher headquarters. This guidance would be used by planners and programmers in developing I&MS capability options. By using the framework as an agreed-on approach for ensuring that their I&MS requirements are met, stakeholders at MAJCOMs and NAFs and higher headquarters can work together with AFIMSC and its detachments located at the MAJCOMs to develop appropriate and consistent objectives for requirements systems,[33] incorporate expeditionary and home-station requirements, and consider both functional and enterprise views, doing so in an iterative fashion that makes the I&MS enterprise agile. AFIMSC will be capable of translating stakeholder objectives into operationally relevant and measurable capabilities (for example, type and number of bases that I&MS resources can support).

This framework can drive the establishment and coordination of modeling assumptions and planning factors in an integrated view within and across career fields composing the I&MS enterprise. By clarifying roles and envisioning analytic capabilities, the framework can enhance

---

[33] See Mills et al., 2014, which highlighted that the current manpower system does not explicitly use expeditionary demands to size and shape all ACS forces.

the ability of AFIMSC to balance competing demands. Implementing this framework can improve the articulation of demands to better communicate the effects of I&MS capability shortfalls.

The feedback loop in the framework can ensure that AFIMSC maintains pace with the changing security environment. While the demand-side stakeholders are contemplating adjustment to OPLANs and force structure to meet future threats, the CS community can provide options for consideration and gain stakeholder buy-in on the resulting I&MS force structure throughout the process. The feedback loop can also drive adjustments to planning factors; processes; and tactics, techniques, and procedures, as well as evaluate the effect of those adjustments on operational support and programming.

4. Implementing the Proposed Framework

AFIMSC needs three enabling mechanisms to make the proposed framework and associated process in Figure 3.1 work. We address those enabling mechanisms in this chapter and propose approaches for creating them.

First, and as mentioned previously, AFIMSC needs a lexicon of metrics to communicate with its customers. Instead of transmitting requirements in terms of numbers of manpower billets, infrastructure projects, or raw dollars, the lexicon requires that resources be articulated and quantified in terms of outputs—ideally in mission or operational outputs, but some kind of metric that articulates to the customer the significance or contribution of the resources.

Second, business rules are needed to ensure equitable support for activities across the service and the ability to discriminate among unique mission requirements where necessary. Beyond simply using a "peanut butter spread" approach, one set of business rules is to stratify bases (or missions or activities) into levels of priority and allocate shortfalls based on some percentage of requirements. We call this the *fill rate approach*. For example, priority 1 bases might get 90 percent of stated requirements; priority 2 bases might get 80 percent of requirements, and so on. This does provide clear prioritization—higher-priority missions and units receive more resources—but the approach is still disconnected from the customer's actual mission and may not provide what that customer needs in terms of capability to accomplish its mission. For example, 80 percent of stated requirements may not be sufficient to support a priority 2 base (even if it is "fair"), but that is not apparent, because there is no visibility over outputs or level of performance. We argue (and describe more fully below) that AFIMSC needs both elements: output-focused metrics and a transparent prioritization scheme (i.e., business rules).

Finally, the Air Force needs a reporting construct to aggregate performance information to make it digestible by its four-star customers.[34] At the most granular level, individual I&MS activities can be articulated according to activity-specific metrics. For example, fewer personnel in an administrative function might mean longer wait times or fewer hours of service per day. Less infrastructure funding might mean a lower level of condition for a road or facility. Less equipment might mean higher risk (for example, firefighting). But the AFIMSC/CC and the MAJCOM/CCs have neither the time nor the bandwidth to monitor as many metrics as there are unique I&MS activities on Air Force installations. A higher-level reporting construct is needed.

As we surveyed the Air Force, we found that there already exist concepts and capabilities that can be adapted or co-opted for each of the three enablers of the framework we propose. We

---

[34] One of the intents of our framework is to enable high-level discussions of resources and capabilities among the MAJCOM customers and the commander of AFIMSC, in accordance with stated needs from the PAD.

describe those concepts and present them as building blocks that could be useful in supporting AFIMSC's resource allocation cycle.

Output-Focused Metrics

The current system for identifying I&MS requirements is rarely output-focused. In other words, the consequences of different funding levels are not clear. To allocate I&MS funds rationally and transparently, AFIMSC needs, as much as possible, to tie funding levels to outcomes. There is a tool that could better enable that connection.

The joint community and the Air Force have adopted a system for defining standardized levels of installation and mission support. The Air Force version is referred to as Common Output Level Standards (COLS). COLS provide a promising tool set that can be incorporated into the enterprise-wide measurement system. Standard levels of support have been identified for 34 I&MS activities, each with a subordinate set of measures that can be used to define four different levels of support. Level 1 is typically classified as full capability and at the same level as the joint basing COLS level 1. Level 2 is viewed as a slightly reduced standard, level 3 as a more reduced standard, and level 4 as a greatly reduced standard that is substandard but still complies with legal requirements and still supports the operational mission.[35] Some functions have established a further definition of each COLS level, with a breakout of the subprogram or activity measures that are considered to determine the level. For example, for fire emergency services, the civil engineering community considers four key services with associated measures and has established a detailed stratification, shown in Table 4.1. (We list the full scope of COLS in Appendix C.)

In this example of fire and emergency services, C2 ties the level of service to the type of incident to which it could respond. Hazardous material (HAZMAT) operations are tied to the duration for which they can sustain a response. Fire operations (which includes extinguishing fires in structures or aircraft), then, are assigned manpower resource levels to its levels of capability and risk. These elements help communicate the connection between resource inputs and performance outputs in ways that communicate both with those allocating resources and with those receiving various levels of support.

---

[35] U.S. Air Force, *Air Force Common Output Level Standards*, PowerPoint Presentation, Washington, D.C., December 2011.

Table 4.1. COLS Levels and Subordinate Measures for Air Force Fire Emergency Services

COLS Level	Definition	Service Area	Level of Support
COLS 1	Full mission support capability	Command & Control	Provide *full* management support to all FES and C2 capability for level 3 incidents.
		Fire Prevention	Provide *full* inspection/educational services to all customers based on DoDI, AFI, and NFPA standards, frequencies, requests, and mission requirements.
		Fire Operations	Provide *full* core mission response capability within response time. Total 18 firefighters available.
		HAZMAT Operations	Provide installation protective action (defensive and offensive) for 4 hours where at least 15 firefighters are available.
COLS 2	Limited mission support capability	Command & Control	Provide *limited* management support to all FES and C2 level 4 incidents.
		Fire Prevention	Provide *limited* inspection/educational services to high life hazard facilities and occupancies.
		Fire Operations	Provide *limited* core mission response capability within response time standards. Total 12 firefighters available.
		HAZMAT Operations	Provide *limited* installation protective actions; defensive operations expected (2-hour duration).
COLS 3	Critical mission support capability	Command & Control	Provide *critical* level of management support to FES operations *only* and C2 capability for level 5 incident.
		Fire Prevention	Provide *critical* level of inspections/educational services to priority mission facilities *only*.
		Fire Operations	Provide *critical* level of services to core mission capability within response time standards. Total 8 firefighters available.
		HAZMAT Operations	Provide *critical* level of services to installation protective actions; defense action expected (1-hour duration).
COLS 4	Inadequate mission support capability	Command & Control	No management support; C2 delegated to fire operations.
		Fire Prevention	No inspection or educational services provided.
		Fire Operations	Provide *inadequate* core mission response capability within response time standards. No follow-on assignment available.
		HAZMAT Operations	No installation protective actions available; firefighters assist others to isolate the area

SOURCE: Headquarters, Air Force Civil Engineering Support Agency, *Fire Emergency Service (FES) Common Output Level Standards*, presentation, Tyndall Air Force Base, Fla., November 2011.
NOTE: FES = fire emergency services; DoDI = Department of Defense instruction; AFI = Air Force instruction; NFPA = National Fire Protection Association.

Another COLS example is SRM facility recapitalization. COLS metrics for that category are driven by the condition rating of infrastructure assets.[36] Yet another example is SRM facility

---

[36] The metrics use a quality rating (or Q-rating), depicted as a facility conditions index, scaled 0 to 100, for each structural asset that relates to the facilities' restoration and modernization needs. For a more technical definition of *Q-rating*, see U.S. Department of Defense, *DoD Real Property Inventory Data Element Dictionary*, Real Property

sustainment. These COLS metrics specify levels of service based on such outcomes as the amount of manufacturer-recommended preventive maintenance that can be performed and the response time for unscheduled maintenance service, both of which are primarily driven by manning level.

We analyzed recent spending data to see how the current COLS map to spending within AFIMSC's likely funding portfolio.[37] (This mapping can be found in Appendix D.) We found that the current COLS appear to cover the full scope of I&MS responsibilities under AFIMSC. Some COLS and their metrics might not currently be suitable for the purposes this framework proposes. Later we discuss some steps necessary to fully implement our approach.

Ultimately, the Air Force has a solid start at performance-based metrics that can be used as a lexicon for communication between AFIMSC and its customers.

Transparent Business Rules to Guide Resource Allocation

Having received customer requirements, AFIMSC can assess resource allocation options. But to do this assessment rationally and transparently, AFIMSC needs a set of business rules. As we survey the functions under AFIMSC in the PAD and the detailed activities enumerated by COLS, we see a potential opportunity to inform these business rules. Our approach has three elements: separate installation support from mission support activities, classify installations and mission activities to guide resource allocations, and develop business rules to make allocations rationally and transparently.

Separating Installation Support and Mission Support

A key question is how resources will be allocated across installations, as well as within installations and across functions. An Air Force installation is a kind of ecosystem composed of many parts. Most Air Force bases have a generic set of installation functions that help the base run smoothly and support the population, irrespective of what mission might be accomplished there (for example, fighter pilot training, satellite control, or cargo operations). The idea we proffer here is that one way to create a rational, transparent process is to decompose installations into their respective installation support functions and mission support functions and treat each according to different standards.

Installation support focuses on those functions that are critical to supporting an installation and its population, and missions not directly tied to force projection. It is helpful to think of the

---

Information Model, Version 4.0, Washington, D.C., April 22, 2010. The Navy uses a tool suite called Shore Facilities Investment Model, run by Booz Allen Hamilton, which contains a module called Macro-Level Forecasting that forecasts the overall health of the infrastructure system based on various levels and distributions of infrastructure funding. A tool or system like this could be used to forecast the condition of Air Force facilities, given the funding scenarios that AFIMSC considers.

[37] This would be the Air Force Total Ownership Cost (AFTOC) database.

installation as a municipality when considering which functions, activities, and resources would fall into this category. We argue that, for the most part, these functions can be funded and supported at a single global level of service, such that an airman on any base would receive at least comparable, if not identical, level of service.

The following is a sample list of the types of activities by functional area that could be included in the installation support category:

- services: morale, welfare, and recreation (MWR); mortuary affairs; lodging; family support; child care
- security forces: installation defense, information security, corrections, noncombatant evacuation
- civil engineering: municipal facility management, municipal infrastructure
- chaplain services
- contracting
- financial management
- transportation: vehicle maintenance, vehicle operations, traffic management operations.

Many of these activities are essentially municipal functions that could be provided to a single global standard. The default would be to provide the same level of service across base types. There could be a few exceptions to this. Some MWR activities can be provided by other means—e.g., other nearby military installations or local community services. Each base's services could be tailored to its specific conditions. However, this does not represent a significant share of the funding portfolio. Second, defense of the installation as a whole, while not strictly a mission function, may need to be differentiated for some extremely critical mission sets. The point of this division between installation and mission support is not to follow some concept of functional purity but simply to enable resource allocation advocacy to happen with greater ease. Any categorization of activities should simplify, not complicate, those processes.

On the other hand, operational mission support would include activities for which AFIMSC would more often differentiate levels of support, driven by unique operational mission requirements. One reason for this is that operational missions are generally the things in the Air Force that are competing with each other and being prioritized. Further, mission activities more often have unique requirements that would command a different level of support. For example, one is the communications infrastructure for a C2 center or a satellite control center. Another example is the electrical power requirements of an aircraft maintenance hangar compared with a flying-simulator building.

The following is an example of the activities within I&MS functional areas that would be included in the operational mission support category:

- services: flight kitchen
- security forces: flight line protection, weapon system protection
- civil engineering: explosive ordnance disposal, fire protection, mission facility management, management of mission infrastructure (e.g., flight line, runway)

- communications: flight line and air traffic control communications networks, mission facility communications infrastructure
- logistics readiness: deployment management
- transportation: flight line and mission-related vehicle operations and maintenance.

For these, each installation or individual mission would argue for special requirements, why they require a higher COLS level (that is, standard levels of support) than other bases and missions, and what level of resources correlate to each COLS level.

We assessed current COLS categories to see how well they might be parsed as installation or mission functions. Given our definitions of these two categories, and how each COLS activity is defined in Air Force documentation, we used our judgment about where each activity best fit. We found 21 COLS categories that could fit exclusively into the installation support category and two COLS categories that could fit exclusively into the mission support category. The remaining 11 COLS categories had subprogram measures that fell into both installation support and mission support categories. Our review found that the current COLS categories provide a reasonable tool for communicating capability requirements at a macro level, as applied to the demand for installation support capability and mission support capability.

Next, we discuss how classifying installations and mission activities can help guide AFIMSC's business rules.

Taxonomy for Classifying Types of Installations and Missions

Each base has different requirements for installation and mission support. Depending on the types of operational mission supported, a degraded capability can have different effects on the Air Force's core missions and its ability to provide combat-ready forces to CCMDs. To help guide prioritization, we propose a taxonomy for classifying installations based on the missions they support and those missions' contribution to the Air Force's support to COCOMs.

Table 4.2 shows a proposed taxonomy for grouping operational missions.

The level of I&MS support for a given installation or mission type will differ based on mission activity. For example, a strategic nuclear base would presumably require higher than normal security for some areas, while a remotely piloted aircraft base or a satellite control installation would require robust and reliable communications to the installation itself and to key facilities. More specifically, one mission might require high base defense, while another might simply require more security for a particular facility or area (for example, military or civilian personnel, supplies, and equipment to perform preventive or emergency maintenance).

We now provide an example from the Air Force logistics system to illustrate the kind of resource allocation system we envision.

Table 4.2. Illustrative Classification Taxonomy for Different Installation Types Based on Mission

Installation Type	Mission Types	Examples
Type 1	• Forward operation locations • Installations that deploy combat forces • Employ-in-place installations • Strategic nuclear mission installations	JB Langley-Eustis Dover AFB Barksdale AFB Ramstein AFB Osan AB Kadena AB F.E. Warren AFB
Type 2	• Installations that provide direct support to contingency operations (e.g., RPA bases, satellite control installations) • Installations that house higher headquarters and warfighting commands	Beale AFB Schriever AFB
Type 3	• Installations that provide critical support to weapon system availability and force readiness (e.g., ALC's, flight test centers, weapon centers, pilot training bases)	Robins AFB Altus AFB Hill AFB
Type 4	• Installations that host nonwarfighting missions (e.g., nonflying training and education, research and development, acquisition)	JB San Antonio Maxwell AFB Hanscom AB Los Angeles AFB

NOTE: JB = joint base; AFB = Air Force base; AB = air base; RPA = remotely piloted aircraft; ALC = Air Logistics Center.

Execution and Prioritization of Repair Support System

The prioritization scheme used in the Execution and Prioritization of Repair Support System (EXPRESS) provides an example of a high-level scheme that stratifies mission needs but still applies detail and rigor.[38]

Today, AFSC provides support for reparable spare parts to the entire Air Force. A myriad of units across diverse mission sets providing dissimilar combat (and training) effects require responsive spare-parts support, presenting AFSC with a knotty prioritization problem. AFSC's current solution is EXPRESS, which takes as objectives the aircraft availability (AA) required by each flying unit (the demand side). EXPRESS uses as the supply side the current resources available to repair broken spare parts (for example, labor, piece parts, funding).[39] Left to its own

---

[38] Which depot-level reparables get inducted for depot repair is determined by daily computations in EXPRESS, also less commonly known by its data system designator, D087X. EXPRESS is the Air Force's implemented version of the tool DRIVE, which was developed at RAND in the early 1990s. The primary function of EXPRESS is to prioritize repair and determine the distribution of repaired parts to maximize weapon system readiness goals. EXPRESS constrains inductions according to availability of funds, capacity, carcasses, and parts to support the repair. See Air Force Materiel Command Instruction 23-120, *Execution and Prioritization Repair Support System (EXPRESS)*, Wright-Patterson Air Force Base, Ohio: Headquarters, Air Force Materiel Command, May 24, 2006.

[39] Headquarters, AFSC/LGS (Performance Management), at Wright-Patterson Air Force Base, runs EXPRESS daily, producing a global priority list. Then, each ALC (and production groups within) can apply its own constraints.

devices, EXPRESS would optimally allocate these resources against the given demands to maximize AA across these diverse fleets.

But Air Force leadership sought some assurance that EXPRESS would still provide the high-level prioritization it desired (which can be difficult with a "black box" solution such as EXPRESS). That guidance is provided in the form of SPRS. This sequence states that several broad categories of demands (specified by Air Force leadership) will be filled before others. First, mission-capable parts (MICAPs) from deployed or deploying units will be filled (those with a JCS code); then, holes in readiness spares package (RSP) kits from such units; then, MICAPs from non-JCS code units; then their kit holes; then, other units; and so on. EXPRESS's optimization logic works *within* those categories, such that all demands from the first category are always (optimally) filled before any from the second (subject to resource availability), and so on. The advantage of this approach is that it makes clear to all stakeholders what high-level prioritization is being applied. The rules are clear, and expectations are managed.

The SPRS in EXPRESS provide an example of how a complex and difficult resource allocation decision for critical support can be made with a combination of explicit statement of demand-side requirements, sophisticated supply-side modeling and analysis, and high-level prioritization guidance and vetting. In our construct for making near-term resource allocation decisions, the base classification taxonomy provides the high-level prioritization guidance. That same taxonomy could also be applied if the decision construct were also applied to long-term programming decisions within the PPBES.[40]

Performance Monitoring Patterned After Existing Air Force Practices

The third step in the process is to monitor performance and provide feedback. Senior leaders are often bombarded with metrics reports, many times without a way to synthesize and digest all the information they are viewing. We therefore propose the development of a performance-monitoring construct patterned after an existing Air Force system that is as simple and transparent as possible.

Air Force Material Command uses the Weapon System Enterprise Review (WSER) to gain a comprehensive view of the status of weapon systems. The review looks at high-level areas, including performance execution, modernization, product support, and predictive health. Under each of these categories, there are subordinate measures, such as not mission capable for supply, not mission capable for maintenance, aircraft delivery quality, cost, schedule, logistics health assessment, and leading health indicators for engines. The status for each metric is reported and contributes to the overall "health" rating for each weapon system. The WSER includes a set of

---

[40] The UMMIPS system could also serve as a pattern. UMMIPS works by assigning a priority designator based on a force activity designator (FAD) (the military importance of the unit) and the urgency of need (how urgently a unit needs a resource, as determined by the circumstances). See Air Force Pamphlet 23-118, *Logistics Codes Desk Reference*, Washington, D.C.: U.S. Air Force, 2012.

business rules that dictate how the overall rating should be adjusted if some number of subordinate metrics are degraded.[41]

Our approach is an enterprise-wide measurement system that is modeled on the WSER and includes a standardized capability rating structure for installation and mission support capabilities.

Installation and Mission Support Measurement System

We propose the creation of a WSER-like measurement system that considers installation support and mission support as enterprise capabilities composed of I&MS functions. The Installation Support Capability Rating (ISCR) and Mission Support Capability Rating (MSCR) together would provide a comprehensive rating of the functions that are required to produce a full capability as opposed to rating each individual functional area. The system could serve as a widely accepted lexicon for defining and measuring I&MS output capabilities.

The ISCR and MSCR system could function in a manner similar to the WSER, with lower-level measures being combined to provide a grade for installation support and mission support. To measure installation support and mission support independently, the I&MS community must delineate between those functional-area activities that support the installation or base population (or might be considered municipality functions), as well as those activities that directly support the operational mission of the base. In some cases, the task of distinguishing between installation support and mission support will require a categorization of assets on the installation, not just functional activities, based on whether they support the installation or the mission. An example of this is facilities. Some facilities on the installation house operational mission activities, while others house nonmission activities. A degradation of a mission facility is more critical than a nonmission facility. Similarly, communications infrastructure and vehicles that support operational mission execution are more critical than those resources that simply support the base's general population. In these cases, the assets would need to be classified and measured separately.

As mentioned, the ISCR would be composed of several I&MS functional areas in whole or in part. The ISCR can be computed from a set of existing metrics used by functional areas today as a part of the COLS system. Appendix C contains a list of COLS and their associated subprogram measures and parses them by their applicability as a measure for the ISCR and the MSCR. Those metrics could roll up to determine the location's ISCR and provide an indication of how well I&MS functions at the base are supporting the installation population and nonoperational mission activities.

Similar to the ISCR, the MSCR is a comprehensive rolled-up measure of I&MS functional areas' ability to support the operational mission. Subordinate-level metrics for I&MS areas that

---

[41] Headquarters, Air Force Material Command, *Weapon System Enterprise Review: Business Rules*, Wright-Patterson Air Force Base, Ohio, June 2015.

are directly tied to the operational mission would be linked through business rules to create the MSCR.

Enterprise-Wide Capability Ratings and Business Rules

The summary rating levels for an installation's ISCR and MSCR could be determined using a set of business rules that dictate when the overall rating is downgraded.[42] The criteria for different MSCR levels can be similar in nature to those for the ISCR but might also be tied to operational mission effect and risk associated with operational mission execution. Table 4.3 provides a notional example of those types of business rules.

Table 4.3. Notional Business Rules for Determining an ISCR and an MSCR

ISCR/MSCR Level	Business Rule	Mission Status
Level 1	100% of functional areas operating at COLS level 1	Fully mission capable
Level 2	10% of functional areas operating at COLS level 2 or 3	Degraded capability but no mission impact (i.e., mitigation sufficient)
Level 3	20% of functional areas operating at COLS level 3 or level 4	Degraded capability and serious mission impact (i.e., mitigation insufficient)
Level 4	More than 20% of functional areas operating at COLS level 3 or lower	Degraded capability and severe mission impact

NOTE: These are simply notional suggestions; the Air Force should consider a more comprehensive set of rules determining the ISCR and the MSCR, as it has done for the WSER.

In addition to the business rules for the ISCR and the MSCR described in Table 4.3, the reporting mechanism could have space for a commander's assessment. This could provide a voice for the local commander—irrespective of the formal metric—to articulate his or her ability to accomplish the mission. This could reveal occasions where the formal metrics do not adequately reflect the way mission support needs to be provided or an opportunity to lower standards, where lower levels of performance are found to be adequate to support a particular mission type.

This proposal will need to be calibrated with customer inputs and insights from subject-matter experts, but one of the keys we highlight here is a distinction we make between situations where local actions can fully mitigate the effects of degraded I&MS capability (level 2) and those situations where local actions are insufficient to mitigate mission impacts (level 3). Below we discuss some of these potential mitigation actions (local commanders have quite a bit of leeway and flexibility to use their resources creatively), but that distinction is important

---

[42] The Air Force uses a comprehensive set of business rules as part of the WSER. For each category measured in the WSER, the business rules specify under what conditions the category should be rated a certain color (green, yellow, red, or blue).

information to be transmitting to AFIMSC and the MAJCOM/CCs to inform dialogue and decisionmaking.

Establishing these business rules before implementing the proposed measurement system and vetting them with stakeholders is a critical step in implementation. The business rules could explicitly state what constitutes a degraded capability, and thus a lower rating, and can serve to establish a degree of consistency across bases reporting their capabilities. Our view is that this system would provide the primary communications mechanism and lexicon for the AFIMSC's engagement with its stakeholder. Similar to the way that the WSER is viewed, a degraded capability sends a strong signal to the AFIMSC that action is needed. As a part of the reporting system, each installation reporting a COLS level lower than acceptable would also be required to report a mitigation plan and associate costs with the mitigation plan.

Enterprise Dashboard and Mitigation Options

Much like the WSER, the ISCR and the MSCR provide AFIMSC an enterprise-wide view of installation and mission support capabilities within the Air Force. The system also provides local installation commanders and MAJCOMs a mechanism for communicating capability status and resource needs to AFIMSC. Following the process outlined in Figure 3.1, the ISCR/MSCR reporting tool can include mitigation options when performance of the system fails to meet the designed or funded output level. Figure 4.1 is a notional WSER-like dashboard for I&MS that shows how the lower-level activities and metrics are rolled up to higher-level summary categories.

Figure 4.1. A Notional I&MS Enterprise Dashboard Modeled After the WSER

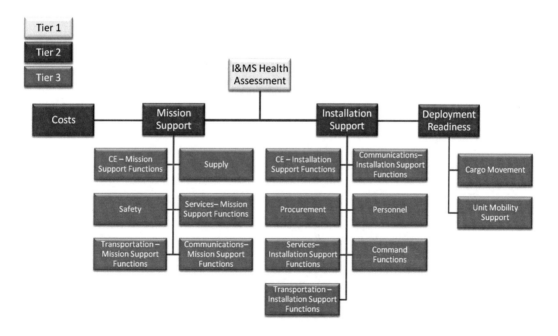

NOTE: CE = civil engineering.

32

The notional I&MS enterprise dashboard reflected in Figure 4.1 could effectively align specific COLS categories under each of the tier 3 categories. To implement this reporting structure, the Air Force could consider an approach similar to that used with the WSER. Generally speaking, each program office reports on a regular basis. The reports contain a summary view of the status of the weapon system, which is developed based on a series of lower-level measures and business rules that dictate when a particular area of performance should be downgraded. The reports include a forecast or prediction of future capability. The WSER also prompts program offices to address mitigation plans and associated costs for items that fall below an acceptable level. We propose a similar approach with the I&MS enterprise dashboard and ISCR and MSCR reporting. Such an approach gives local commanders, who should have at first pass the best insights for mitigating a shortfall, a means of communicating with AFIMSC. For example, local commanders have several choices about how to meet their mission with reduced resources. While AFIMSC will allocate resources to installations based on a rough expectation of a level of service, local commanders actually have quite a bit of leeway and flexibility to use those allocated resources as they see fit to support their missions. The authority of AFIMSC should not limit those local commanders' ability to do so.

For reduced manpower levels, a local commander generally has four choices:

- **work personnel harder**: extending work hours of remaining personnel
- **reduce services**: temporarily reducing services to mitigate the shortfall; this could simply entail taking more risk, depending on the function
- **reallocate personnel temporarily**: this could mean putting personnel on guard duty, or supporting "extra" duties out of hide (for example, honor guard).
- **backfill personnel temporarily**: during deployments, funds may also be available (for example, existing operations and maintenance funds or in some cases overseas contingency operations (overseas contingency operation funding) to temporarily backfill personnel while they are deployed.

If infrastructure SRM funding is reduced, one major effect is to delay projects that would improve the condition of a facility or other asset—for example, replacing a roof. Delaying such a project increases the workload on organic personnel to temporarily troubleshoot and patch problems that arise in the intervening period before that project can be accomplished. This is the type of mitigation we envision will be captured in this performance-monitoring system.

Contingency Support

Thus far, we have dealt with installation requirements mostly in the absence of contingency-driven inputs. We now examine two ways in which AFIMSC resource allocations can and must keep in mind contingency requirements.

First, some of the installations in the AFIMSC portfolio are forward bases that could directly engage in warfighting (for example, Korea). This priority should be reflected in the relevant prioritization and allocation business rules, as discussed above. Second, AFIMSC will provide POM inputs via the ACS CFL to allocate military manpower authorizations to all bases. The personnel who fill these spots will mostly support day-to-day training and other activities on their respective installations, but together they contribute to a global pool of capability that will deploy to forward bases (or support employ-in-place missions). Plausibly, AFIMSC could propose a force-shaping action that would reduce military authorizations in one area for economic reasons that could inadvertently compromise warfighting capability and capacity. This is one area where the expeditionary arm of AFIMSC can provide warfighting requirements to act as a constraint in allocating manpower authorizations that may be seen as primarily a peacetime resource. Past RAND research presents approaches for allocating ACS manpower authorizations in ways that integrate expeditionary and home-station requirements.[43]

Methodology for Implementing the Measurement System

Implementing the measurement system will not be easy. To that end, we propose the approach outlined in Table 4.4.

Table 4.4. Step-Wise Approach for Implementing an Enterprise-Wide Measurement System

Step	Purpose	Status
1. Categorize I&MS functions	Delineate between installation support and mission support activities	Proposed breakout presented in Appendix C
2. Validate demand drivers	Validate the demand drivers by functions and catalog the expected output or production value	To be accomplished; can use allowance standards as starting point
3. Establish output-level categories	Develop output levels for each function and its associated production value or risk	Current COLS is a major starting point
4. Map existing resources	Align resources at each installation to its appropriate installation support or mission support category	To be accomplished by each installation
5. Implement enterprise measures	Establish metrics hierarchy Establish business rules for summary-level measures	Need validation of hierarchy and business rules
6. Evaluate shortfalls	Assess ISCR and MSCR ratings of each installation, impact on mission, and the cost of mitigating shortfalls	Would be accomplished as part of regular review cycle
7. Allocate resources	Consider affected mission types from base classification taxonomy to allocate resources	Example of taxonomy proposed in Chapter Three

---

[43] See, for example, Mills, Drew, et al., 2014; and Albert A. Robbert, Lisa M. Harrington, Tara L. Terry, and Hugh G. Massey, *Air Force Manpower Requirements and Component Mix: A Focus on Agile Combat Support*, Santa Monica, Calif.: RAND Corporation, RR-617-AF, 2014.

The left column in Table 4.4 highlights the key steps to implementation. The middle column illuminates those steps and provides a more complete description. The right column provides a view of where the Air Force has already made progress toward implementation.

Fortunately, the Air Force has many of the pieces in place. For example, the creation of AFIMSC as an agency to oversee and manage the I&MS enterprise creates an entity with the overall responsibility to execute the steps in Table 4.4. The agency consolidates responsibilities in a single organization and gives the entire I&MS community a unified presence at the table. Step 3, which is to establish output-level categories and associated measures, is also well under way. The Air Force's progress with COLS is a major step that is critical to implementing this system.

Similarly, there is great progress on step 5. Other measurement systems in place today within Air Force Materiel Command, such as the WSER, provide AFIMSC a strong set of policy and procedures as examples from which to govern the ISCR/MSCR system, as well as examples of business rules that can be used to govern the rating system. Finally, as a part of this analysis, we provide a first cut at step 1 in Appendix C, where we show how COLS measures could be aligned under the two top rating categories, and step 7, where we provide a notional taxonomy for categorizing bases to assist with prioritizing resource allocation decisions across competing demands. However, there is more work to be done to implement this measurement system. Specifically, the following need to occur:

- There is a need for a more detailed review of current COLS categories and how they can be aligned into either the ISCR or the MSCR. The proposed alignment was based on COLS categories and subprogram measures that were originally approved by the COLS Executive Steering Group. It is likely that some of those have been refined as the system has been executed.
- The current COLS should be evaluated to ensure they provide a comprehensive view of installation support capability and/or mission support capability requirements. For example, the Executive Steering Group–approved COLS for security forces explicitly state that they do not apply to weapon system security, which may well be a large portion of security forces requirements.
- The linkages between COLS capability levels as a demand driver and the rule sets for the resources required to meet the COLS levels must be further examined. This refers to the shortfall mentioned earlier—different functional areas use different sets of rules to establish resource requirements. Being able to closely link resource requirements to COLS levels is a fundamental aspect of this proposed measurement system and decision framework. This is described as step 2 in the implementation steps outlined in Table 4.4.
- Similar to those used in the WSER, there must be business rules to determine what triggers a downgrade in the overarching ISCR or MSCR. The business rules for the ISCR and the MSCR must be considered independently for each rating based on risk and the cost of failure.

35

Integrating the ISCR/MSCR Measurement System into Long-Term Planning Across the PPBES Horizon

The approach proposed here for near-term allocation decisions also supports long-term resource planning and allocation decisions. During the programming phase of the PPBES process, manpower and equipment standards and infrastructure SRM funds would be aggregated into a total I&MS requirement. Individual bases would help translate their unique requirements into COLS levels so the funding requirements could be communicated in terms of local performance, and vice versa.

An enterprise-wide decision could be made to fund installation support and mission support at particular levels. Such a decision can serve as the baseline for subsequent program requests from the ACS CFL with inputs from AFIMSC. AFIMSC, working in conjunction with installations, could make a POM submission based on the resources needed to achieve the respective level.

The POM build process could iterate with trade-offs between COLS levels and across installation categories until an acceptable budget request is developed for the entire I&MS enterprise and incorporated into the ACS CFL input. For example, the initial POM build might use as a baseline the assumption that all installations will be funded at a MSCR COLS level 1 and an ISCR COLS level 2. If, after the initial review of the proposed POM input, the requirement exceeds a reasonable level that the I&MS enterprise could expect to compete for within the Air Force Corporate Structure, a second iteration might include a modification where only certain installation types will be funded at an MSCR COLS level 1, while all other installation types are programmed at an MSCR COLS level 2.

Using this approach in the PPBES process provides several key benefits to the Air Force generally and the AFIMSC specifically. First, the process is inclusive and provides a mechanism for installations to make inputs. Second, it considers functional areas as capabilities composed of complementary, mutually exclusive, completely exhaustive components, in much the same way that Major Force Programs are viewed and compete within the PPBES process. Third, it serves as a mechanism to set expectations for the level of support that can be achieved at any given installation during the year of execution. Finally, after Congress approves the budget and funds are distributed, the process establishes a baseline against which units can be measured during the year of execution. In addition, this process incorporates the enterprise view of resource allocation that the Air Force sought in establishing AFIMSC and assigning this new organization with that goal.

During the year of execution, reporting through the ISCR/MSCR system will reveal whether an installation's capability is below or above the level for which it is budgeted. If a unit is only funded to a COLS level 3 and it reports a COLS level 3, then there is no need for rebalancing or redistributing resources by the AFIMSC. If a unit is funded for COLS level 2 but it is operating at a COLS level 3, its unfunded requirements would be a candidate for additional funding. The same methodology could be used for making capability cuts when there are budget constraints. A deliberate and transparent decision could be made about where to allocate capability cuts and the cost savings that would be expected as a result of making those cuts.

5. Conclusions, Recommendations, and Next Steps

The analysis outlined in the previous chapters leads us to conclusions about the analytical framework needed by AFIMSC. Those conclusions, along with specific recommendations and next steps, are outlined in this chapter.

Conclusions

As a result of the current fiscal and security environments, as well as the role of AFIMSC as the enterprise manager of I&MS capabilities, evolving the analytical capabilities of AFIMSC is a complex challenge. Our analysis leads us to the following conclusions:

- As the principal supplier of installation and mission support resources, AFIMSC needs a coherent, rational, and transparent method to allocate resources across missions and installations.
- AFIMSC cannot just adopt the status quo system where resource allocations were made absent enterprise-wide standards for support and insights into the operational impact of those decisions.
- A common lexicon of metrics, clear business rules, and a construct to report enterprise performance to senior leaders is needed to implement a framework for making rational resource allocation decisions and have the ability to communicate the impact of those decisions to operators.
- Some processes that the Air Force has already established can be tailored to facilitate creating an enterprise measurement and management structure. Air Force Material Command's WSER, which provides a comprehensive review of the status of weapon systems, is an example.

Recommendations

To mitigate the challenges of the current security and fiscal environments, and to capitalize on the transformative initiatives affecting the ACS community, the Air Force needs to:

- Implement an analytical framework for AFIMSC, as the supplier of I&MS resources, that provides meaningful trade-off information to both the customers of I&MS capabilities and an integrator responsible for adjudicating mismatches between demand and supply.
- Develop, in concert with and agreed on by the operations community, a lexicon and set of business rules to inform the decision tradespace of the operations community.
- Implement an approach to an enterprise-wide measure system that is based on the WSER, in use at Air Force Materiel Command today, and includes a standardized capability rating structure for installation and mission support capabilities.
- In implementing that system, the Air Force should employ a step-wise approach.

- Incorporate the ACS planning, execution, monitoring, and control framework and associated process into the broader Air Force C2 system and use the integrated system to present meaningful trade-offs to the combatant commanders and a DoD integrator.

Next Steps

Defining the analytical framework is just the beginning, and the framework described in this document only addresses the doctrine and process components of the DOTmLPF-P construct.[44] The additional elements must be considered to achieve a fully functional analysis capability as AFIMSC progresses toward full operational capability. For example, the following questions need to be answered:

- What is the full suite of tools and analytic capabilities needed to enable examination of the appropriate tradespaces across all assigned ACS resources?
- What processes are needed to enable close collaboration by the Expeditionary and Installation directorates to ensure inputs required by the Integration Directorate are available?
- What types of personnel, training, and education are needed to facilitate the analysis of demands and options for supplying demands across all assigned ACS resources?
- What types of organizational interfaces and reporting constructs will be required with MAJCOMs, C-NAFs, Air Staff, and other organizations to facilitate the understanding of potential means of satisfying demands across all assigned ACS resources?
- What are the barriers to implementing such a framework?

---

[44] The DOTmLPF-P construct is Doctrine, Organization, Training, materiel, Leadership and education, Personnel, Facilities, and Policy (Air Force Policy Directive 10-6, *Capability Requirements Development*, Washington, D.C.: U.S. Air Force, November 6, 2013).

Appendix A. Summary of Prior RAND Research on ACS C2

The creation of AFIMSC has links to prior RAND research focused on ACS activities of planning, execution, monitoring, and control. The Air Force adopted several of the recommendations made by RAND in its body of ACS research that subsequently evolved to the creation of such organizations as AFSC and AFIMSC. It is important to understand the prior ACS research associated with C2,[45] because it is a foundation to the recommendations proposed here for the analytical framework for AFIMSC.

Chapter Two highlights the original body of research on C2 of combat support, including the creation of an operational architecture. Over the years, follow-on analyses further expanded and updated the architecture, addressed implementation actions, and proposed additional recommendations for enhancing ACS processes. Those analyses are documented in the following reports:

- Patrick Mills, Ken Evers, Donna Kinlin, and Robert S. Tripp, *Supporting Air and Space Expeditionary Forces: Expanded Operational Architecture for Combat Support Execution Planning and Control*, Santa Monica, Calif.: RAND Corporation, MG-316-AF, 2006. This report expands and provides more detail on several organizational nodes in our earlier work that outlined concepts for an operational architecture for guiding the development of the Air Force combat support execution planning and control needed to enable rapid deployment and employment of the Air and Space Expeditionary Force. These combat support planning, execution, and control processes are sometimes referred to as ACS C2 processes.

- Robert S. Tripp, Kristin F. Lynch, John G. Drew, and Robert DeFeo, *Improving Air Force Command and Control Through Enhanced Agile Combat Support Planning, Execution, Monitoring, and Control Processes*, Santa Monica, Calif.: RAND Corporation, MG-1070-AF, 2012. This report compares the current state of ACS planning, executing, monitoring, and controlling with the suggested implementation actions designed to address shortfalls identified in the 2002 RAND Project AIR FORCE operational architecture. The report further recommended implementation strategies to facilitate changes needed to improve Air Force C2 through enhanced ACS planning, executing, monitoring, and control processes.

- Kristin F. Lynch and William A. Williams, *Combat Support Execution Planning and Control: An Assessment of Initial Implementations in Air Force Exercises*, Santa Monica,

---

[45] C2 of Air Force forces is accomplished by planning, executing, monitoring, and controlling the application of capabilities. The Air Force and joint communities have corporate C2 systems to accomplish those activities. The planning, executing, monitoring, and controlling of ACS resources is accomplished within the context of the Air Force and joint systems. Prior RAND research used a variety of terms to describe planning, executing, monitoring, and controlling the application of ACS capabilities, including *execution planning and control*; *planning, execution, and control*; *command and control*; and *ACS C2*. These terms all refer to the application of ACS capabilities to support operational needs within the larger Air Force and joint C2 systems.

Calif.: RAND Corporation, TR-356-AF, 2009. This report evaluates the progress the Air Force has made in implementing the TO-BE ACS operational architecture as observed during the operational-level C2 warfighter exercises Terminal Fury 2004 and Austere Challenge 2004 and identifies areas that need to be strengthened. By monitoring ACS processes, such as the development of combat support requirements for force package options that were needed to achieve desired operational effects, assessments were made about organizational structure, systems and tools, and training and education.

- Kristin F. Lynch, John G. Drew, Robert S. Tripp, Daniel M. Romano, Jin Woo Yi, and Amy L. Maletic, *Implementation Actions for Improving Air Force Command and Control Through Enhanced Agile Combat Support Planning, Execution, Monitoring, and Control Processes*, Santa Monica, Calif.: RAND Corporation, RR-259-AF, 2014. This report identifies and describes where shortfalls or major gaps exist between current ACS processes and the vision for integrating enhanced ACS processes into Air Force C2. The report evaluates C2 nodes from the level of the President and Secretary of Defense to the units and sources of supply. It also evaluates these nodes across operational phases and suggests mitigation strategies needed to facilitate an efficient and effective global C2 network.

- Kristin F. Lynch, John G. Drew, Robert S. Tripp, Daniel M. Romano, Jin Woo Yi, and Amy L. Maletic, *An Operational Architecture for Improving Air Force Command and Control Through Enhanced Agile Combat Support Planning, Execution, Monitoring, and Control Processes*, Santa Monica, Calif.: RAND Corporation, RR-261-AF, 2014. This report presents an architecture that depicts how enhanced ACS processes could be integrated into Air Force C2, as it is defined in joint publications. This architecture, which focuses on the near term (the next four to five years) using current Air Force assets, was created by (1) evaluating previous RAND-developed operational architectures from 2002 and 2006 and (2) refining those architectures in light of the current operational and fiscal environments. It first identifies C2 processes and the echelons of command responsible for executing those processes and then describes how enhanced ACS planning, execution, monitoring, and control processes could be integrated with operational-level and strategic-level C2 processes to provide senior leaders with enterprise ACS capability and constraint information.

Each of the above analyses built on the original ACS operational architecture and addressed additional shortfalls that were exposed in contingencies such as Operations Enduring Freedom and Iraqi Freedom. The operational architecture was revised and updated as additional shortfalls were observed, new security environments emerged, and Air Force processes evolved.

Prior Research Identified Several Shortfalls Within ACS Planning, Execution, Monitoring, and Control

The analyses we have cited in this appendix—drawing on interviews with Air Force personnel and observation of contingencies and exercises—revealed a number of shortfalls within Air Force ACS planning, execution, monitoring, and control processes. The shortfalls were categorized into several themes that were then addressed by the operational architecture and recommendations for implementing the architecture.

Independent C2 processes: A fundamental shortfall was inconsistent and incomplete integration of ACS planning, execution, monitoring, and control processes within the Air Force and joint operations C2 processes and systems. The chasm between combat support and operations was accentuated by neither community having a comprehensive understanding of the planning and C2 activities of the other. Operators used terms such as *operational effects* and *sortie rates* to describe their measures of success, and logisticians would use such terms as *inventory quantities, mission-capable rates, days of supply,* and *distribution response time* to describe the capabilities of the ACS enterprise. The ACS measures were not meaningful to the operators who needed to know how many sorties they could generate in the coming days.

Absence of performance-feedback loops: In addition to the disconnected processes, the ACS planning, execution, monitoring, and control processes lacked feedback loops that could provide insights into ACS enterprise performance relative to what that performance needs to be to achieve operational effects desired by the warfighting commanders. The absence of the feedback loop limited the ability to dynamically and quickly adjust the ACS plan during execution. Operators would often plan to generate a certain number of sorties unaware that, as a result of prior-day operations, there were insufficient ACS resources to support these plans.

Lack of demand arbitration: Related to both of the previous shortfalls is the absence of a commonly accepted method and set of business rules to arbitrate the allocation of resources when there are competing demands. In some cases, those conducting simultaneous operations in different theaters would unknowingly plan to use the same ACS resources, such as munitions, and then have to adjust their plans at the last moment when the ACS network discovered the overlapping demand.

Additional shortfalls, such as the evolution of C2 systems to support the fundamental process and training and education for ACS personnel, were identified in the prior research, but those shortfalls and associated mitigation recommendations do not lend value to the historical perspective as related to the evolution of AFIMSC.

RAND-Proposed Recommendations to Mitigate ACS Planning, Execution, Monitoring, and Control Shortfalls

The prior RAND analyses on enhanced ACS processes recommended actions to mitigate the identified shortfalls. The proposed process and organizational relations were documented in the original operational architecture, which was expanded and refined over the course of more than ten years. The expansions included the addition of more detail to demonstrate the applicability of the operational architecture to PPBES activities and updated process definitions to match both the changing operational environment and Air Force structure.

In addition to developing the operational architecture, RAND proposed a set of actions to implement it. Table A.1 summarizes the recommendations that addressed doctrine, processes, organization, training and education, and tools and systems.

Table A.1. Steps to Improve ACS Planning, Execution, Monitoring, and Control

To Achieve This Goal	Take This Action
Enhance processes	Focus ACS planning, execution, monitoring, and control process on operational outcomes; identify and separate supply, demand, and integrator processes; include closed-loop feedback and control
Expand doctrine	Delineate roles of ACS nodes, including logistics, operational, and installation staff; Air Force commanders; MAJCOMs, the Air Force Global Logistics Support Center; and others
Refine training and expand education	Educate Air Force staff officers in ACS planning and staff responsibilities and strategies-to-tasks methodology; assign some promotable "supply-side" officers to "demand-side" organizations and vice versa
Implement systems and tools	Identify critical ACS communications and information system capabilities needed to assess, monitor, and inform allocation decisions and update as necessary
Strengthen organizations and instructions	Assign supply, demand, and integrator processes to organizations and functions; modify instructions and other documents to support ACS assessment and control functions

SOURCE: Tripp, Lynch, Drew, and DeFeo, 2012.

The organization recommendations have been adopted by the Air Force in some fashion and over time evolved to the creation of AFSC and AFIMSC. Understanding the evolution of those organizations in the context of the broader Air Force ACS planning, execution, monitoring, and control framework is an important underpinning of the recommendations we provide here for AFIMSC's analytical framework.

The Evolution of Air Force ACS Planning, Execution, Monitoring, and Control Organizations

The 2002 RAND study on ACS planning, execution, monitoring, and control processes recommended the creation of centralized organizations to integrate the ACS enterprise and provide a single interface to operational C2 organizations.[46] Two organizations in particular, inventory control points (ICPs) and a global integration center (GIC), have links to what has evolved into AFIMSC.

ICPs were designed to provide an integrated view of supply capabilities for each commodity.[47] The idea was that the ICPs for spares "would manage the spares along a continuum of operations, with immediate access to both the data and analytical tools needed to assess capability and manage distribution of resources to MAJCOMs and theaters under direction from an integrating function."[48] The integrating function would exist within the GIC, which was viewed as "a virtual organization with cells co-located with Air Combat Command (ACC), Air

---

[46] Leftwich et al., 2002, p. 44.

[47] Although ICPs were largely focused on the management and control of materiel resources, AFIMSC focuses on both materiel and personnel resources.

[48] Leftwich et al., 2002, p. 48.

Mobility Command (AMC), and [Air Force] Space Command [AFSPC] Regional Support Squadrons (RSSs) to assess weapon system capabilities and should have responsibility for providing integrated weapon system assessments across commodities both in peacetime and wartime."[49] A combination of the ICPs and the GIC would provide an enterprise view of ACS capabilities and the ability of the ACS enterprise to respond to operational demands.

The Air Force created a GIC-type of organization for ACS resources focused on mission generation when it created the Air Force Global Logistics Support Center (AFGLSC). Over time, the AFGLSC functions were integrated into the AFSC as part of the Air Force Materiel Command transformation.

A RAND report by Lynch et al. (2014) expanded the 2002, 2006, and 2009 studies and addressed the expanded scope of ACS to include installation support functions. That analysis produced an OA view that focused on establishing, sustaining, and protecting the base. In the spirit of the GIC, which was proposed in the original 2002 study, and the creation of the AFGLSC for mission-generation functions, the 2014 report proposed the creation of a global installation manager that would fulfill the GIC function specific to installation support activities.

---

[49] Leftwich et al., 2002, p. 51.

Appendix B. Quantifying the Effects of Home-Station ACS Shortfalls

The Air Force needs the ability to assess the impacts of deployments on home-station operations. Prior RAND work developed a method to quantify the impacts of deployments on home-station operations, in a way that has the potential to inform resource trade-offs and policymaking.[50] A significant number and proportion of many home-station services are provided by active duty personnel in ACS career fields, along with civilians and contractors. Some of these career fields are sized only to meet peacetime home-station requirements, so when active duty forces deploy, home-station bases must somehow mitigate this loss of personnel.

Commanders at the wing level and below have four choices to mitigate these personnel shortfalls. They can require active duty personnel to work longer hours, reduce on-base services, use operations and maintenance funding to temporarily backfill the losses, or temporarily delegate tasks to personnel or units that would not normally provide them. At any given base, some combination of these is usually implemented; the decision driven by local conditions, commander judgments of impact and risk, and availability of funding. The selected course of action could have negative impacts on military personnel (and by extension career field health), the overall base population, training, and other home-station activities (for example, employ-in-place missions).

Our analysis of this challenge considered a suite of metrics to help inform a range of Air Force decisions, especially force structure shaping and reductions and risk assessments. The metrics were the following:

- **Effective workload per person.** This estimates the net workload increase on active duty personnel remaining at home station during a deployment, based on a range of factors specific to each career field and base.
- **Service reduction.** This is usually career-field specific and quantifies the amount of services reduced in terms specific to how each career field's manpower requirements are initially generated.
- **Backfill funding requirements.** This estimates the cost of providing reservist, civilian, or contractor personnel on a temporary basis to backfill active duty deployments.
- **Training load.** This is the ratio of E-3 enlisted personnel to E-5 enlisted personnel. This quantifies the on-the-job-training burden on senior enlisted personnel to train junior enlisted personnel. The burden often increases at home station during deployments, because more senior personnel are often deployed preferentially.

---

[50] This was unpublished research conducted in fiscal year 2012 for a project titled "Analytic Support to the ACS Core Function Lead Integrator (CFLI)."

- *Deploy-to-dwell ratio.* This is not a new metric, but we recommend retaining it in the suite of metrics the Air Force uses. This is the ratio of time spent deployed (deploy) to time spent at home (dwell). This also sometimes called the *spin rate.*

To assess these metrics quantitatively, we analyzed existing Air Force data and information in the form of manpower-requirement documents (known as *manpower standards*), historical deployments, historical and current manpower and personnel levels, and interviews with subject-matter experts.

The following is a summary of our findings:

- The Air Force currently has manning shortfalls and skill and grade imbalances; deployments can exacerbate the challenges that the shortfalls and imbalances induce.
- Most career fields do not experience workload reductions during deployment, even those whose workloads are primarily population-driven.
- For career fields that do experience workload reductions during deployment, the extent of those reductions can be quantified using manpower standards.
- For career fields that do experience workload reductions during deployment, workload reductions are not proportional to base-population reductions because career field-sizing rules often incorporate fixed and variable components. The impact of deployments on each of our proposed metrics can be quantified using historical data and future scenario data using current Air Force deployment concepts.
- Deployments affect our proposed metrics to different extents, depending on a range of variables. For example, a deployment that greatly increases workload per person might not affect training load.
- These metrics can be used to set policies, shape the force, and assess risk to home-station operations.

In this analysis, we used manpower standards to develop the mathematical relationships between home-station manpower drivers and resulting manpower requirements. An example of these relationships shows that as a base population increases, so does its services manpower requirement, and vice versa. Thus, as personnel deploy from a base, the services manpower requirement naturally declines. But services personnel also deploy, so there is a net shortfall because service personnel deploy faster than their workload decreases.

Figure B.1 shows the home-station trade-offs for the services career field across the entire active duty Air Force. The x-axis shows the number of deployed services personnel; the y-axis shows the resulting aggregate services personnel shortfall at active duty home-stations. This shortfall is shown as the purple line in the figure. At zero personnel deployed, the services career field already has a shortfall relative to requirements (260 personnel). This is because it is undermanned relative to validated requirements (by about 6 percent for active duty). Moving left to right, as services personnel (and their respective populations) deploy, the shortfall increases. This shortfall may be mitigated in the ways described above.

Figure B.1. Quantifying the Impacts of Home-Station ACS Shortfalls

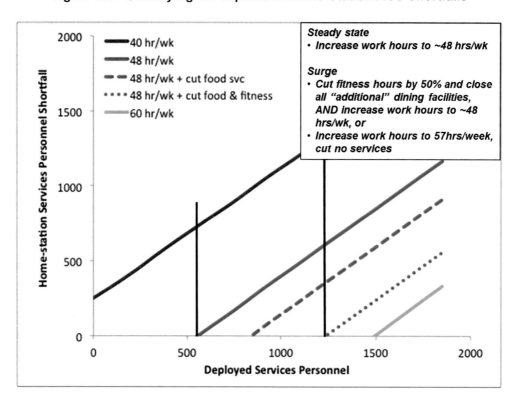

NOTE: The leftmost black vertical line shows the *average* active duty services personnel deployed from 2006 to 2011, and the rightmost black vertical line shows the *maximum* active duty services personnel deployed since 2001.

In the figure, we illustrate two mitigation options: extending work hours and reducing services. The solid green and orange lines show the net shortfall if active duty work hours are extended to 48 or 60 hours per week, respectively. The green dashed line shows the shortfall if the work hours are 48 hours per week *and* all additional dining facilities are temporarily closed. The green dotted line shows the results if fitness centers hours are reduced by 50 percent *in addition to* the reductions indicated by the green dashed line. Finally, the leftmost black vertical line shows the *average* active duty services personnel deployed from 2006 to 2011, and the rightmost black vertical line shows the *maximum* active duty services personnel deployed since 2001. At either deployed level, the resulting shortfall could be mitigated by working longer, reducing services, or some combination. Alternatively, the cost to mitigate the restricted fitness center hours by half (to lessen the impact to home station) at the average deployed level using reservists would be about $1 million per month.

Similar calculations can be done for planning scenarios to estimate the impacts and trade-offs of future operations. RAND has also analyzed several other ACS career fields and has developed a method that could be applicable across the Air Force. These assessments potentially can be used to inform force-shaping, risk assessment, resource allocation, and unit type code availability coding. In the context of AFIMSC, there will be near-term installation support challenges

47

associated with the current operations tempo and deployment rates to support ongoing contingencies. This analysis can assist AFIMSC in establishing policies for home-station levels of support given the current operations tempo.

Appendix C. Illustrative Alignment of COLS to Installation Support and Mission Support Capability Ratings

Table C.1 reflects an illustrative parsing of COLS and subprogram measures into the installation support category or a mission support category. The idea presented here is that within each functional area managed by AFIMSC, there are subordinate activities, as well as subprograms within those activities. Within the COLS hierarchy, many of those subprograms have specific output functions that are measured. In some cases, the output function ties directly to providing support for the installation population (thinking of the installation as a municipality); in other cases, the output function ties directly to supporting the operational mission hosted on the installation. In a few cases, the delineation is not so clear by activity and will require a mapping of assets on the installation to determine whether they are most closely aligned to installation population support or to operational mission support. An example of that type of mapping was developed in unpublished RAND analysis for linking infrastructure resources to readiness.

Table C.1. Illustrative Alignment of COLS Metrics to Installation and Mission Support Categories

COLS Category	Subprogram/Metrics	Installation Support	Mission Support
Big Three	Custodial Service	X	
	Grounds Maintenance	X	
	Integrated Solid Waste (Refuse & Recycling)	X	
Demolition	Air Force Goal	X	
	Facility Demolition	X	
Emergency Management	Planning		X
	Preparedness		X
	Operations		X
Environmental	Compliance	X	
	Conservation	X	
	Pollution Prevention	X	
Facility Recapitalization	Facility Condition	X	X
Facility Sustainment	Execution of Distributed Funding	X	
	Unscheduled Maintenance	X	X
	Scheduled Preventive Maintenance (PM)	X	X
	Scheduled Component Replacement & Repair	X	X
Fire Emergency Services	Command & Control		X

COLS Category	Subprogram/Metrics	Installation Support	Mission Support
	Fire Prevention		X
	Fire Operations		X
	HAZMAT Operations		X
Housing Mgt	Family Housing Overseas	X	
	Oversight of Privatized Housing	X	
	Unaccompanied Housing	X	
Pavement Clearance	Snow and Ice Control: Priority One: Red Areas		X
	Snow and Ice Control: Priority Two: Yellow Areas		X
	Snow and Ice Control: Priority Three: Green Areas	X	
	Airfield Sweeping Operations		X
	Roadway Sweeping Operations	X	
Pest Management	Indoor Pest Management	X	
	Outdoor Pest Management	X	
Real Property and Engineering Services	Real Estate/Property Asset Management	X	
	Facility and Infrastructure Maintenance Repair and Construction Projects Program	X	X
	Space Utilization	X	
	Service Contract Quality Assurance	X	
	Installation Master Planning	X	
Utilities	Purchased Energy (Commodities)	X	X
	Central Plant Operations	X	X
	Purchased Potable Water & Waste Water Removal	X	
	Treatment Plant Operations Potable Water & Waste Water Removal	X	
Communications Command Support	Records Mgt, Privacy, FOIA	X	
	Postal	X	
Data Transmission Services	Collaboration and Messaging Services	X	X
	Fixed Voice	X	X
	Help Desk Support	X	X
	Information Assurance	X	X
	Infrastructure Support	X	X
	Network Availability	X	X

COLS Category	Subprogram/Metrics	Installation Support	Mission Support
	Wireless Connectivity	X	X
	Video Telecommunications	X	
Procurement Ops	Procurement Acquisition Lead Time (simplified)	X	
	Procurement Acquisition Lead Time (large acq)	X	
	GPC Program Management	X	
Civilian Personnel	Staffing	X	
	Classification	X	
	Employee Mgmt Relations	X	
	Labor Mgmt Relations	X	
	Benefits & Entitlements (B&E)	X	
	Systems Support	X	
Equal Opportunity	Education and Training	X	
	Military Complaint and Incident Processing	X	
	Civilian Complaint Processing	X	
	Organizational Assessments and Briefings	X	
Financial Management	Non-appropriated Funds Financial Analysis	X	
	Quality Assurance (Compliance)	X	
	Financial Systems Support (LAN)	X	
	Financial Analysis	X	
	Financial Services	X	
Chapel Ministries	Religious Functions	X	
	Unit Ministry	X	
	Marriage and Family Care	X	
	Special Events	X	
Honor Guard	Military Funeral Honors (MFH)	X	
	Honors at Arrival Airports	X	
	Honor Guard Training	X	
	Official and Civic Event Support	X	
Inspector General	Investigations	X	
	Assistance	X	
	Training	X	
Legal Support	Military Justice	X	
	Administrative Law	X	
	Claims	X	

COLS Category	Subprogram/Metrics	Installation Support	Mission Support
	Environmental Law	X	
	Contract Law	X	
	Labor Law	X	
	International & Operations Law	X	
	Legal Assistance	X	
Military Personnel	Customer Support	X	
	Career Development	X	
	Force Management	X	
	IT Systems Support	X	
	Installation Personnel Readiness	X	
Public Affairs	Communication Counsel	X	
	PA Outreach	X	
	PA Operations	X	
	Visual Information	X	
Supply Storage	Requisition	X	X
	Inventory Management	X	X
	Reutilization of Materiel, Products and Customer Returns		
	Store and Warehouse	X	X
Safety	Training	X	
	Inspections and Evaluations	X	X
	Mishap Investigations	X	X
	Safety Awareness and Campaigns	X	
Security Forces	Installation Entry Control (IEC)	X	
	Commercial Vehicle Inspection (CVI)	X	
	Incident Response	X	
Child and Youth Programs	Child Care	X	
	Youth Programs	X	
Food Services/Laundry	Dining Facilities	X	
	Flight Kitchen		X
	Operational Rations		X
	Laundry	X	
	Dry Cleaning	X	
Lodging	Customer Perspective	X	
	Program Capacity	X	
MWR	Fitness	X	

COLS Category	Subprogram/Metrics	Installation Support	Mission Support
	Library	X	
	Community Centers	X	
Warfighter and Family Services	Deployment Readiness		X
	Personal and Family Life Readiness	X	
	Economic Readiness	X	
	Information and Referral Services	X	
Base Support Vehicles and Equipment	Air Force Owned Vehicle/Vehicular Equipment Maintenance		X
	Provide Transportation Support Services	X	
	Provide Class-C Pooled Vehicle Support	X	
Installation Movement	Official Passenger Travel	X	
	Personal Property Moves	X	
	Cargo Movement Services		X
	Unit Mobility Support		X
	Manage Installation Transportation Office	X	

Appendix D. Illustrative Mapping of Program Element Categories to Air Force COLS

Figure D.1 contains an illustrative mapping of program element categories (PECs) to COLS.

Figure D.1. Illustrative Mapping of Program Element Code Categories to COLS

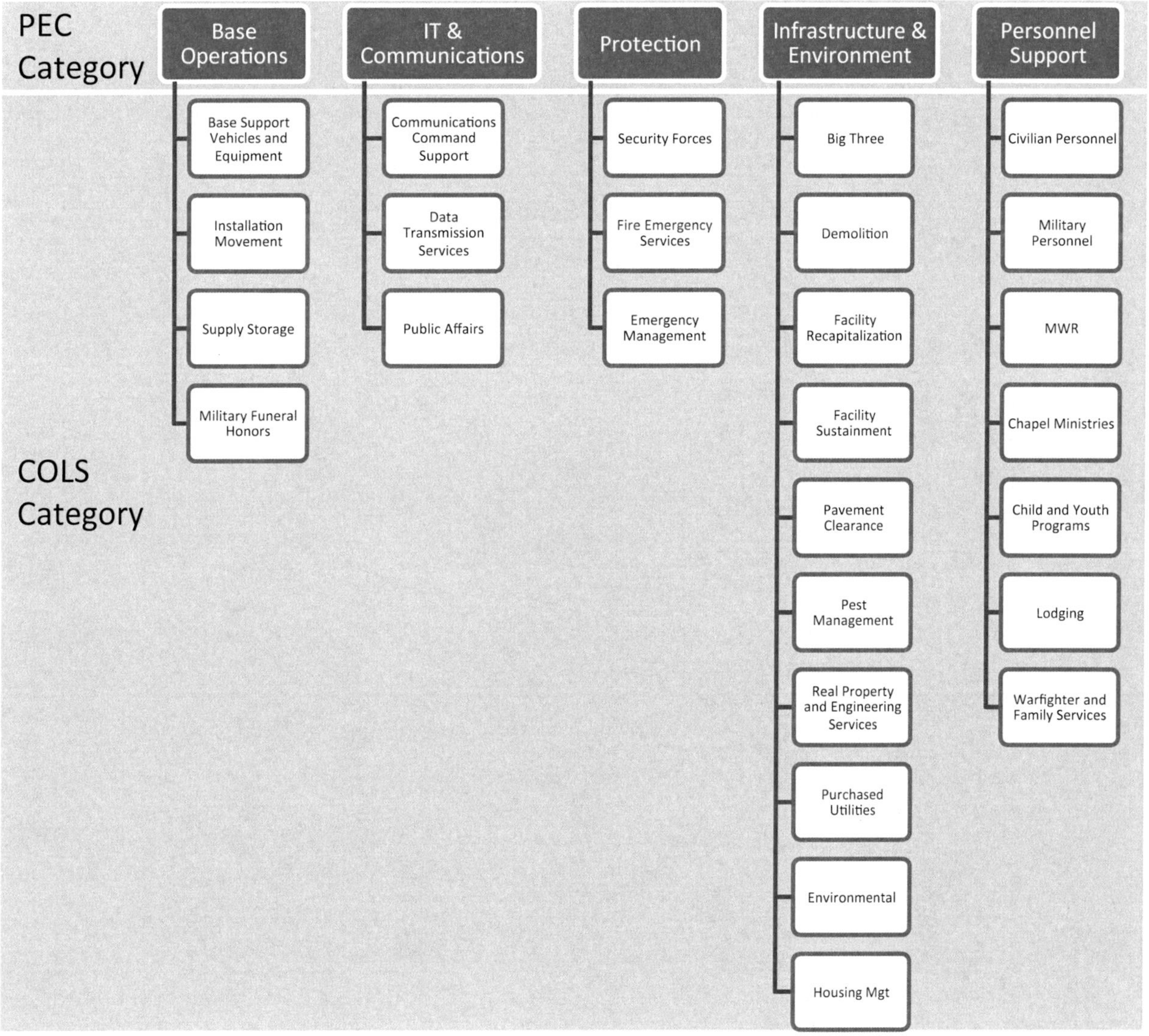

PEC Category	Base Operations	IT & Communications	Protection	Infrastructure & Environment	Personnel Support
	Base Support Vehicles and Equipment	Communications Command Support	Security Forces	Big Three	Civilian Personnel
	Installation Movement	Data Transmission Services	Fire Emergency Services	Demolition	Military Personnel
	Supply Storage	Public Affairs	Emergency Management	Facility Recapitalization	MWR
COLS Category	Military Funeral Honors			Facility Sustainment	Chapel Ministries
				Pavement Clearance	Child and Youth Programs
				Pest Management	Lodging
				Real Property and Engineering Services	Warfighter and Family Services
				Purchased Utilities	
				Environmental	
				Housing Mgt	

References

Abell, John, Louis W. Miller, Curtis E. Neumann, and Judith E. Payne, *DRIVE (Distribution and Repair in Variable Environments): Enhancing the Responsiveness of Depot Repair*, Santa Monica, Calif.: RAND Corporation, R-3888-AF, 1992. As of September 4, 2016:
http://www.rand.org/pubs/reports/R3888.html

Air Force Materiel Command Instruction 23-120, *Execution and Prioritization Repair Support System (EXPRESS)*, Wright-Patterson Air Force Base, Ohio: Headquarters, Air Force Materiel Command, May 24, 2006.

Air Force Pamphlet 23-118, *Logistics Codes Desk Reference*, Washington, D.C.: U.S. Air Force, 2012.

Air Force Policy Directive 10-6, *Capability Requirements Development*, Washington, D.C.: U.S. Air Force, November 6, 2013.

Drew, John G., Ronald G. McGarvey, and Peter Buryk, *Enabling Early Sustainment Decisions: Application to F-35 Depot-Level Maintenance*, Santa Monica, Calif.: RAND Corporation, RR-397-AF, 2013. As of August 25, 2016:
http://www.rand.org/pubs/research_reports/RR397.html

Headquarters, Air Force Civil Engineering Support Agency, *Fire Emergency Service (FES) Common Output Level Standards*, presentation, Tyndall Air Force Base, Fla., November 2011.

Headquarters, Air Force Material Command, *Weapon System Enterprise Review: Business Rules*, Wright-Patterson Air Force Base, Ohio, June 2015.

Jensen, Michael C., and William H. Meckling, "Specific and General Knowledge and Organizational Structure," in Lars Werin and Hans Wijkander, eds., *Contract Economics*, Oxford: Basil Blackwell, 1992.

Leftwich, James A., Robert S. Tripp, Amanda B. Geller, Patrick Mills, Tom LaTourrette, Charles Robert Roll, Jr., Cauley von Hoffman, and David Johansen, *Supporting Expeditionary Aerospace Forces: An Operational Architecture for Combat Support Execution Planning and Control*, Santa Monica, Calif.: RAND Corporation, MR-1536-AF, 2002. As of September 4, 2016:
http://www.rand.org/pubs/monograph_reports/MR1536.html

Lewis, Leslie, James A. Coggin, and C. Robert Roll, *The United States Special Operations Command Resource Management Process: An Application of the Strategy-to-Tasks Framework*, Santa Monica, Calif.: RAND Corporation, MR-445-A/SOCOM, 1994. As of September 4, 2016:
http://www.rand.org/pubs/monograph_reports/MR445.html

Lynch, Kristin F., John G. Drew, Robert S. Tripp, and Charles Robert Roll, Jr., *Supporting Air and Space Expeditionary Forces: Lessons from Operation Iraqi Freedom*, Santa Monica, Calif.: RAND Corporation, MG-193-AF, 2005. As of September 4, 2016:
http://www.rand.org/pubs/monographs/MG193.html

Lynch, Kristin F., John G. Drew, Robert S. Tripp, Daniel M. Romano, Jin Woo Yi, and Amy L. Maletic, *Implementation Actions for Improving Air Force Command and Control Through Enhanced Agile Combat Support Planning, Execution, Monitoring, and Control Processes*, Santa Monica, Calif.: RAND Corporation, RR-259-AF, 2014a. As of September 4, 2016:
http://www.rand.org/pubs/research_reports/RR259.html

———, *An Operational Architecture for Improving Air Force Command and Control Through Enhanced Agile Combat Support Planning, Execution, Monitoring, and Control Processes*, Santa Monica, Calif.: RAND Corporation, RR-261-AF, 2014b. As of September 27, 2016:
http://www.rand.org/pubs/research_reports/RR261.html

Lynch, Kristin F., and William A. Williams, *Combat Support Execution Planning and Control: An Assessment of Initial Implementations in Air Force Exercises*, Santa Monica, Calif.: RAND Corporation, TR-356-AF, 2009. As of September 29, 2016:
http://www.rand.org/pubs/technical_reports/TR356.html

Mills, Patrick, John G. Drew, John A. Ausink, Daniel M. Romano, and Rachel Costello, *Balancing Agile Combat Support Manpower to Better Meet the Future Security Environment*, Santa Monica, Calif.: RAND Corporation, RR-337-AF, 2014. As of September 4, 2016:
http://www.rand.org/pubs/research_reports/RR337.html

Mills, Patrick, Ken Evers, Donna Kinlin, and Robert S. Tripp, *Supporting Air and Space Expeditionary Forces: Expanded Operational Architecture for Combat Support Execution Planning and Control*, Santa Monica, Calif.: RAND Corporation, MG-316-AF, 2006. As of September 29, 2016:
http://www.rand.org/pubs/monographs/MG316.html

PAD—*See* Program Action Directive.

Program Action Directive 14-04, *Implementation of the Air Force Installation and Mission Support Center (AFIMSC)*, Washington, D.C.: Headquarters, U.S. Air Force, December 2014.

Robbert, Albert A., Lisa M. Harrington, Tara L. Terry, and Hugh G. Massey, *Air Force Manpower Requirements and Component Mix: A Focus on Agile Combat Support*, Santa Monica, Calif.: RAND Corporation, RR-617-AF, 2014. As of September 4, 2016: http://www.rand.org/pubs/research_reports/RR617.html

Tripp, Robert S., John G. Drew, and Kristin F. Lynch, *A Conceptual Framework for More Effectively Integrating Combat Support Capabilities Constraints into Contingency Strategy Development and Execution*, Santa Monica, Calif.: RAND Corporation, RR-1025-AF, 2015. As of September 4, 2016: http://www.rand.org/pubs/research_reports/RR1025.html

Tripp, Robert S., Kristin F. Lynch, John G. Drew, and Robert G. DeFeo, *Improving Air Force Command and Control Through Enhanced Agile Combat Support Planning, Execution, Monitoring, and Control Processes*, Santa Monica, Calif.: RAND Corporation, MG-1070-AF, 2012. As of September 4, 2016: http://www.rand.org/pubs/monographs/MG1070.html

U.S. Air Force, *Manpower Standard: Combat Support Flight*, AFMS 45XA, Washington, D.C., March 30, 2005.

———, *Air Force Common Output Level Standards*, PowerPoint presentation, Washington, D.C., December 2011.

———, *Weapon System Enterprise Review Business Rules*, Washington, D.C., June 2015.

U.S. Department of Defense, *DoD Real Property Inventory Data Element Dictionary*, Real Property Information Model, Version 4.0, Washington, D.C., April 22, 2010.